Public Perception of Climate Change

Despite the findings on global climate change presented by the scientific community, there remains a significant gap between its recommendations and the actions of the public and policy makers. So far, scientists and the media have failed to successfully communicate the urgency of the climate change situation in such a way that long-term, comprehensive, and legally binding policy commitments are being made on the national and international level. This book examines the way the public processes information, and how they perceive threats and other perceptual factors that have a significant effect on how and to what degree climate change mitigation and adaptation strategies are supported.

Understanding public risk perception plays a vital role in communicating the challenges of global climate change. Using a diverse range of international case studies, this book explores the nature of public perceptions of climate change and identifies the perception factors which have a significant impact on the public's willingness to support global climate change policies or commit to behavioral changes to reduce greenhouse gas emissions and improve urban resiliency. The comparative study of social and cultural factors, beliefs, attitudes, and trust provides an international overview of best practices regarding the design, implementation, and generation of public support for climate change policies at a global level.

Offering valuable insight into climate change and risk communication, the book will be of interest to students and scholars of environment studies, politics, urban planning, and media and cultural studies.

Bjoern Hagen is Assistant Research Professor at Arizona State University, USA.

Routledge Studies in Environmental Communication and Media

Culture, Development and Petroleum
An Ethnography of the High North
Edited by Jan-Oddvar Sørnes, Larry Browning and Jan Terje Henriksen

Discourses of Global Climate Change
Apocalyptic framing and political antagonisms
Jonas Anshelm and Martin Hultman

The Troubled Rhetoric and Communication of Climate Change
The argumentative situation
Philip Eubanks

Environmental Communication and Travel Journalism
Consumerism, Conflict and Concern
Lynette McGaurr

Environmental Ethics and Film
Pat Brereton

Environmental Crises in Central Asia
From steppes to seas, from deserts to glaciers
Edited by Eric Freedman and Mark Neuzil

Environmental Advertising in China and the USA
Structures of desire
Xinghua Li

Public Perception of Climate Change
Policy and communication
Bjoern Hagen

Public Perception of Climate Change

Policy and communication

Bjoern Hagen

Routledge
Taylor & Francis Group

LONDON AND NEW YORK

First published 2016
by Routledge

2 Park Square, Milton Park, Abingdon, Oxon OX14 4RN
711 Third Avenue, New York, NY 10017, USA

Routledge is an imprint of the Taylor & Francis Group, an informa business

First issued in paperback 2017

British Library Cataloguing-in-Publication Data
A catalogue record for this book is available from the British Library

Library of Congress Cataloging-in-Publication Data
A catalog record for this book has been requested.

ISBN: 978-1-138-79523-5 (hbk)
ISBN: 978-1-138-10425-9 (pbk)

Typeset in Bembo
by Apex CoVantage, LLC

Contents

Figures

Tables

Preface

Climate change is among the most important issues of the 21st century. Adaptation to and mitigation of climate change are some of the salient local and regional challenges scientists, decision makers, and the general public face today and will be in the near future. However, designed adaptation and mitigation strategies do not guarantee success in coping with global climate change. Despite the robust and convincing body for anthropogenic global climate change research and science, there is still a significant gap between the recommendations provided by the scientific community and the actual actions by the public and policy makers.

In order to design, implement, and generate sufficient public support for policies and planning interventions at the national and international level, it is necessary to have a good understanding of the public's perceptions regarding climate change. Based on the ongoing "Global Survey on Public Attitudes towards Climate Change" research project, the purpose of the study presented in this publication is two-fold: First, to understand the nature of public perceptions of global climate change in different countries; and second, to identify perception factors which have a significant impact on the public's willingness to support climate change policies or commit to behavioral changes to reduce greenhouse gas emissions. Factors such as trust in climate change information, which need to be considered in future climate change communication efforts, are also dealt with in this dissertation.

This study has identified several aspects that need to be considered in future communication programs. Climate change is characterized by high uncertainties, unfamiliar risks, and other characteristics of hazards which make personal connections, responsibility, and engagement difficult. Communication efforts need to acknowledge these obstacles, build up trust and motivate the public to be more engaged in reducing climate change by emphasizing the multiple benefits of many policies outside of just reducing climate change. Levels of skepticism among the public towards the reality of climate change as well as the trustworthiness and sufficiency of the scientific findings varies by country. Thus, communicators need to be aware of their audience in order to decide how educational their program needs to be.

The first chapter provides an overview of the issue of climate change by presenting the scientific basis of climate change and the role of the Intergovernmental Panel on Climate Change (IPCC), current climate change impacts, future impacts of climate change, and the uncertainty in climate change projections. Furthermore, the chapter discusses the current shift in focus by policy makers form mitigation to adaption as one of the main response strategies of climate policy. The last part of this introductory chapter points to the duality of urbanities as both contributors and victims of climate extremes as well as emphasizes the important role of planners and decision makers in reducing climate change and its impacts.

Chapter 2 provides an overview and best practices of adaptation and mitigation policies and strategies, especially from an urban planning perspective. The existing literature suggests that planning strategies have great potential in achieving significant greenhouse gas (GHG) emission reduction as well as decreasing vulnerabilities and increasing adaptive capacities of urban environments. Barriers and shortcomings are discussed and linked to the important role of public perceptions and improved communication between scientists, policy makers, and other decision makers, as well as the lay public. Finally, the chapter concludes with emphasizing the important role of understanding public climate change perceptions for successful climate change communication efforts and policy implantations.

Chapter 3 introduces the "Global Survey on Public Attitudes towards Climate Change" research project started in 2010. The research project included a nine-country survey of public perceptions and attitudes in 2010–2011. This chapter consists of two main parts. The first part discusses the research methodology utilized in the research project, especially the survey component. The second part is a literature review providing an overview of the body of knowledge in the areas of risk perception and communication relevant to the issues of climate change.

Chapter 4 takes a close look at how the survey participants, from the nine countries, responded to questions dealing with the potential threats and risks of climate change, the saliency of the issue, trust in climate change information and their sources, and preferences for specific policies and strategies. Results of key survey questions are compared among the countries identifying differences and similarities as well as relationships between important variables. Since policy makers do not have a good understanding of where various publics stand on climate change strategies and what will be acceptable and supportive, the survey data assist policy makers in evaluating the appropriate choices to make and determine what would be seen as publically acceptable decision making.

The fifth chapter focuses on the advanced data analysis of the "Global Survey on Public Attitudes towards Climate Change" research project. The analysis identifies perception factors which have a significant impact on the public's willingness to support climate change. In addition, the chapter covers the role of socioeconomic variables, personal knowledge, as well as perceived trust and sense of responsibility towards potential risk communicators. The importance of "trust" is further underlined by illustrating the relationship between public trust in the science of climate change and the different sources of information and risk perceptions of climate change on the global scale. This is an emerging area in climate change science and little understood until now.

The final Chapter 6 presents a discussion of the research results and places the insights and conclusions gained into the context of existing literature as well as with the underlying theories, hypothesis, and research questions of this study. The final part of this chapter addresses the meaning and implication of the knowledge gained for future communication efforts, emphasizes country-specific differences in the findings, and points to future research questions. In addition, a framework for decision makers is presented to make better informed decisions developing, communicating, and implementing appropriate climate change policies that are supported by the public and therefore can be more likely to be implemented successfully.

Acknowledgments

The work presented in this book would not be possible without the essential support of the Foundation of Innovation (Stiftung für Innovation) Rhineland-Palatinate, Germany, and Arizona State University's (ASU) School of Geographical Sciences and Urban Planning, as well as ASU's Lightworks Initiative.

1 Introduction to climate change

Climate change: an overview

Climate change is one of the most important science and societal issues of the 21st century. Although climate change is often perceived as a global issue, impacts can already be observed in both the national and local scales (Pittock, 2009; National Research Council [NRC], 2010). This global trend does not exclude the developed world and rich countries such as the United States. Extreme weather events are increasing in frequency, affecting different regions and sectors throughout the country. Areas are facing climate conditions that have never been experienced before. Among other impacts, ongoing drought and increases in temperatures in the southwest United States have led to an earlier start of, and longer lasting, wildfire season. Prolonged droughts have also increased the competition for limited water resources among people and eco-systems. In other regions such as the Northeast, Midwest, or the Great Plains, weather and climate data show that over the past century, heavy rainfalls, which exceed the capacity of infrastructure systems such as storm drains and sewer systems, have increased. This has led to an uptake in flooding events, land erosion, and landslides (Intergovernmental Panel on Climate Change [IPCC], 2013).

The majority of the scientific community is in agreement that human behavior and current urban patterns, especially automobile usage, are key factors in the rapid increase of the average global temperature in recent decades (Calthorpe, 2010). Cities cover less than 1 percent of the earth's surface, but are disproportionately responsible for causing greenhouse gas (GHG) emissions and are the major cause of climate change. Most of the world's energy consumption either occurs in cities or is a direct result of the way cities function. Cities are not only a major contributor to climate change, but they also play a major part in solving current and future challenges stemming from climate change. Urban sustainability plays a key role in the discussion about the causes, impacts, and solutions of climate change. In recent years, there has been growing acknowledgement among scientists and policy makers that the challenges of climate change can be met through the design, development, and redesign of urban space and structure. Making cities more sustainable will not only reduce the causes of climate change (mitigation) but also improve our resiliency towards the effects of climate change that cannot be avoided anymore (adaptation).

Climate is the most significant component of the world as we know it. The landscape, plants, and animals are greatly influenced by long-term climate conditions, and urban areas are often affected by short-term climate fluctuations. Prior to the introduction of irrigation and the start of industrialization, climate determined food supplies, trade, trade routes, and where people could live. In general, the term *climate* is the typical range of weather and its variability experienced at a particular place (Archer & Rahmstorf, 2010). People speak of *climate variability* describing the irregularities of weather at a particular location from one year (or decade) to another. Changes over longer time scales are referred to as *climate change*. Although, modern technology allows people to live in places where it was impossible to live earlier, local climate is still a key factor for choosing an appropriate design for buildings and urban areas. Aside from some natural variations, scientists concur that the rapid changes experienced in climate during the last several decades are mostly caused by human activity or are anthropogenic, that is, resulting from the influence of human beings on nature (IPCC, 2013). It is well known that climate can change over time. Yet, the acceleration of the rate of change and the observability of the global warming trend recorded in the last few decades are alarming.

As shown in Figure 1.1, worldwide surface temperatures have increased since 1901. Each of the last three decades has been warmer than any other decade since 1850. Between 1880 and 2012, the averaged combined land and ocean surface temperature has increased by 0.85°C (33.53°F). Simultaneously, climate data indicate that since 1950, the number of unusual and extremely cold days and nights are decreasing, whereas the number of abnormally warm days and nights are becoming more frequent.

Figure 1.2 shows the temperature trends by continent, global averages between 1910 and 2010, and compares these observations with simulated climate change emphasizing the impact of human behavior on the significant increase of temperatures over the last 50 years. The black lines show the actual measured temperatures. The dark grey band shows simulated temperatures from various climate models assuming that only natural causes are impacting climate conditions. The light grey band shows the spread of model outputs due to human actions, such as GHG emissions as causes for climate change. The observed temperatures consistently overlap with the simulated temperatures assuming anthropogenic climate change, suggesting that global temperature trends, especially since 1950, cannot be explained through natural causes alone. Research shows a larger than 95 percent level of confidence (IPCC, 2013) that the majority of the observed increases in global average surface temperatures since 1951 was caused by human-induced GHG emissions. Overall, from 1950 to 2000, the warming trend was around 0.13°C per decade, almost twice as much as in the 100-year period before (IPCC, 2007). This trend continued to the beginning of the 21st century and is expected to accelerate even further in the future. Worldwide, the years between 2000 and 2009 are the warmest decade ever measured (National Aeronautics and Space Administration [NASA], 2010). More recently, the National Oceanic and Atmospheric Administration

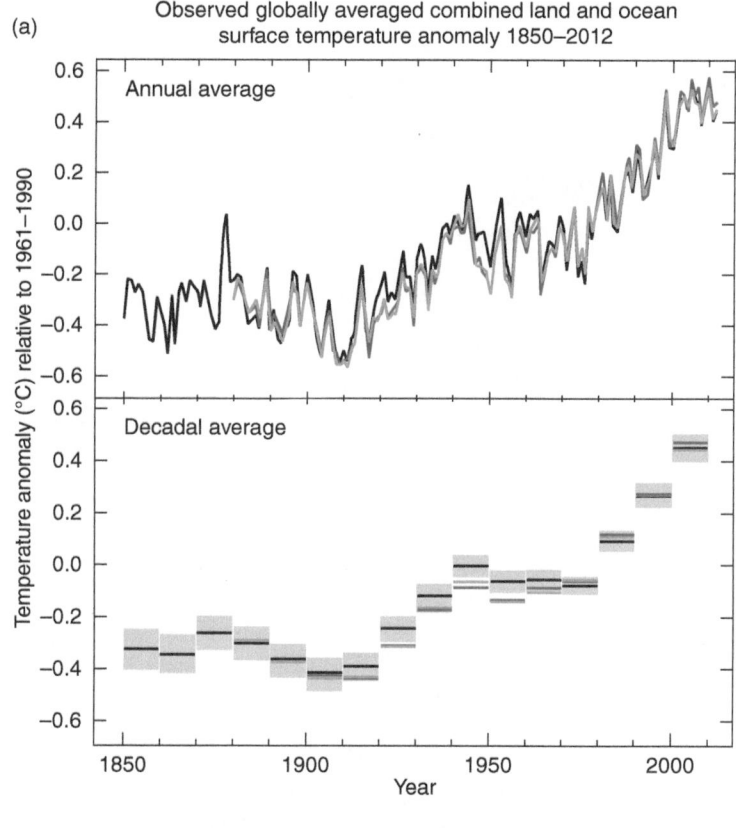

(a) Observed globally averaged combined land and ocean surface temperature anomaly 1850–2012

Annual average

Decadal average

Temperature anomaly (°C) relative to 1961–1990

Year

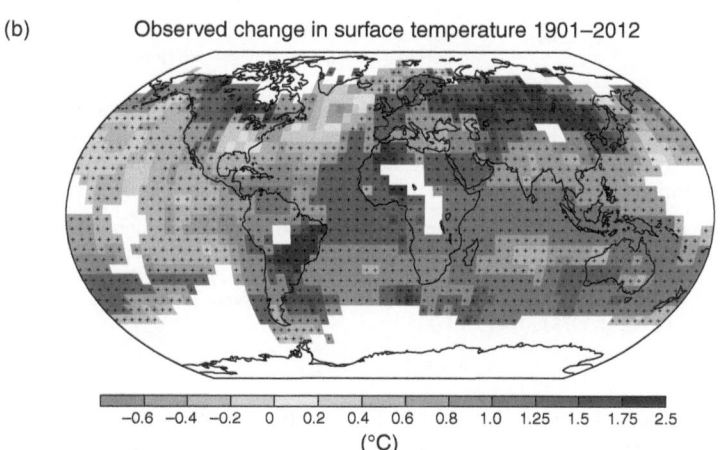

(b) Observed change in surface temperature 1901–2012

−0.6 −0.4 −0.2 0 0.2 0.4 0.6 0.8 1.0 1.25 1.5 1.75 2.5
(°C)

Figure 1.1 Observed annual and decadal global mean surface temperature anomalies from 1850 to 2012 and map of the observed surface temperature change from 1901 to 2012

Source: Intergovernmental Panel on Climate Change (2013). Summary for policymakers. In T.F., Stocker, D. Qin, G.-K. Plattner, M. Tignor, S.K. Allen, J. Boschung, . . . P.M. Midgley (eds.), *Climate change 2013: The physical science basis. Contribution of Working Group I to the Fifth Assessment Report of the Intergovernmental Panel on Climate Change* (pp. 1–30). Cambridge, UK/New York, NY: Cambridge University Press. doi:10.1017/CBO9781107415324.004.

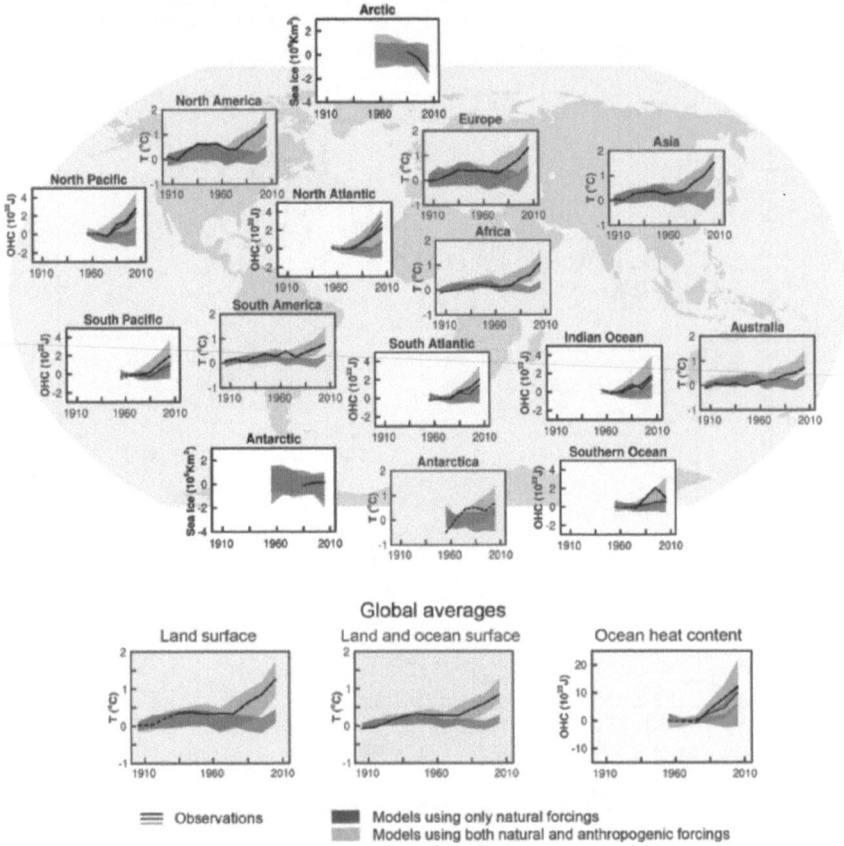

Figure 1.2 Comparison of observed and simulated climate change. The black lines show the actual measured temperatures. The dark grey band shows the expected temperatures if climate would only be influenced by natural causes. The light grey band show possible trends of temperatures if human actions are acknowledged as causes for climate change.

Source: Intergovernmental Panel on Climate Change. (2013). Summary for policymakers. In T.F., Stocker, D. Qin, G.-K. Plattner, M. Tignor, S.K. Allen, J. Boschung, ... P.M. Midgley (eds.), *Climate change 2013: The physical science basis. Contribution of Working Group I to the Fifth Assessment Report of the Intergovernmental Panel on Climate Change* (pp. 1–30). Cambridge, UK/New York, NY: Cambridge University Press. doi:10.1017/CBO9781107415324.004.

(NOAA) concluded that the year 2013 was the fourth-warmest year on record (NOAA, 2013). The records go back to 1880 and show that the current top 10 warmest years all occurred since 1998.

The scientific basis of climate change and the role of the IPCC

Since 1979, when the National Academy of Sciences (NAS) first raised concern about global warming, the body of knowledge and the amount of scientific data observing this phenomena have grown. The year 1988 marked the start of the Intergovernmental Panel on Climate Change (IPCC), founded by the World Meteorological Organization (WMO) and the United Nations Environment Programme (UNEP). Today, the IPCC is considered the leading institution for the assessment of climate change. Its mission is to monitor the scientific work done worldwide regarding climate change. With the help of thousands of scientists, the IPCC regularly assesses the available scientific information relevant for improving the understanding of climate change and its possible environmental and socioeconomic impacts. The participating scientists are divided mainly into three "Working Groups": (1) the scientific assessment of today's research regarding climate change (IPCC, 2013); (2) the potential impacts of climate change to socioeconomic and natural systems and how they can be reduced (IPCC, 2014a); and (3) evaluating options for avoiding the causes of climate change (IPCC, 2014b). The results are summarized and published in specific chapters of the "Assessment Reports of the Intergovernmental Panel on Climate Change."

Since 1988, the IPCC has released five Climate Change Assessment Reports. The first report was released in 1990 and did not find sufficient scientific proof to demonstrate a relationship between human behavior and climate change. Nevertheless, the report projected that by the year 2000, the connection between human actions in the emission of greenhouse gases and climate change would be made. The second report, released in 1995, concluded that evidence of human-induced climate change has already been found, 5 years earlier than projected by the first report. The third report in 2001 supported this claim and provided more evidence that the increase in temperature over the past 50 years can be linked to GHG emissions. In 2007, the fourth report finally stated that (with a likelihood of 90–99 percent) climate change is driven by human-caused emissions of heat-trapping gases and that serious environmental damages can be expected in the future. The reports of the three different working groups for the fifth assessment report were released between September 2013 and April 2014. For scientific reports, the documents use strong language emphasizing the urgency to take action against climate change. The latest assessment report concludes with a 95 percent certainty that human behavior (human-induced greenhouse gas emissions) is the main reason for global warming since 1950.

The basic principle of the earth's climate is that the energy entering the atmosphere from the sun is reflective and has to go out again. The sun's energy is mostly submitted by sunlight either in the form of visible or ultraviolet light. A great proportion of incoming radiation is reflected back to space by snow, ice, and clouds. The sunlight that is reflected back away from the earth is referred to as the earth's albedo and does not "deposit" energy. If this steady exchange between incoming and outgoing energy is out of balance, meaning that less energy is being reflected back to space than in the past, then the temperature from the earth's surface and atmosphere changes. The lower the albedo, the greater the heat absorption is on earth. Various factors that can change the earth's temperature are called climate forcing agents; and the strength of these factors is called radiative forcing.

The human-caused emissions of greenhouse gases have played a key role in the changes of the earth's temperature in recent decades (greenhouse effect). Greenhouse gases such as CO_2 and methane (CH4) function as heat-trapping gases, preventing the earth's albedo from reflecting the sun's energy back into space. Instead, those gases trap the energy inside the earth's atmosphere and send parts of it back to the surface. As a result, over time, the atmosphere and the earth surfaces warm up more and cool down less. The four most common GHG released by humans are carbon dioxide (CO_2), methane (CH4), halocarbons, and nitrous oxide (N2O). Overall, GHG emissions have increased 70 percent between 1970 and 2004, with carbon dioxide being the largest contributor (IPCC, 2007). Systematic measurements of the concentration of carbon dioxide in the earth's atmosphere began in the 1950s. Since then, the concentration of CO_2 has been rising at an accelerating rate. For example, the concentration of CO_2 rose 20 percent faster between the years 2000 to 2004 compared to the 1990s. The majority of the emitted CO_2 has been captured and stored by the ocean, resulting in increased levels of sea water acidification. In addition to its high percentage in the earth's atmosphere, CO_2 also has a longer lifetime than many other gasses emitted by humans. Its persistency factor is quite high.

Greater CO_2 concentration in the atmosphere is, to a large extent, due to an increase in human caused CO_2 emissions from fossil fuel combustion, deforestation, and cement manufacture. Burning of fossil fuels, mostly by private automobiles, is the largest single source of CO_2 emissions (IPCC, 2013). The transport sector alone accounts for about 23 percent of the overall CO_2 emissions from fossil fuel combustion (International Transport Forum, 2010). Another source of significant CO_2 emissions is the building sector. In the United States, buildings account for approximately 40 percent of the total CO_2 emissions (USGBC, 2012). This large amount of emissions by the building sector is due to their high electricity consumption and the fact that much of that energy is created by the burning of fossil fuels, such as coal or natural gas. Deforestation causes high CO_2 emissions from burning or decomposing of trees and soil carbon.

Methane is the second most frequent GHG found. Compared to a molecule of CO_2, the radiative forcing from a molecule of methane is about 30 times stronger. However, with a current lifetime of about 8 years, methane molecules lifespan is much shorter than CO_2. Another difference with CO_2 is that the concentration of methane in the atmosphere has not increased since 1993 (Archer & Rahmstorf, 2010). Natural and artificial wetlands, as well as oil wells, are the largest sources of methane emissions. Although methane concentrations are currently stable, methane sources are expected to increase due to thawing permafrost, another example of climate change impacts. Halocarbons and nitrous oxide have a significantly smaller impact on climate change than carbon dioxide and methane. The concentrations of halocarbons in the atmosphere, however, are declining as a result of international efforts to protect the ozone layer (NOAA, 2005).

Current climate change impacts

An average temperature increase of 0.85°C (33.53°F) since the beginning of the 20th century might not sound like much. However, the impacts of climate change are already visible in the United States and globally. Increases in air and water temperature reduced frost days. A higher frequency and magnitude of heavy rainfall, a rise in sea level, reduced snow cover, glaciers, permafrost, and sea ice are also observed. These changes can affect human health, water supply, agriculture, coastal areas, and the natural environment. One recent conclusion is that in many areas of the world, climate change impacts are occurring faster than once expected (Pittock, 2009).

Sea level rise

The increase in ocean temperatures and the melting of ice sheets are both direct results of climate change and the main contributors to ocean expansion and the *rise of the sea level*, observed since the beginning of the 20th century. The process of ocean water expanding as it gets warmer is called *thermal expansion* and has a relatively small effect. Considering that the oceans are on average 3,800 meters deep, even an average expansion of one hundredths of 1 percent would result in the ocean rising by 38 centimeters, which poses a significant risk for coastal cities and its inhabitants.

The possible melting of ice sheets is a far bigger issue. If all ice sheets melted entirely, the sea level could rise by as much as 70 meters, which would change the world's coastal landscapes forever. Currently, data show that the sea level has risen 19 centimeters since the beginning of the 20th century (IPCC, 2013). More important, recent studies summarized in the latest IPCC reports strongly suggest that the rate of sea level rise has accelerated since 1901. Between 1901 and 2010, the global average sea level rise was 1.7 mm per year, the annual average for the time period between 1971 and 2010 was 2.0 mm, and since 1993, the annual average increase in sea level ranges from 2.8 to 3.6 mm. The two ice sheets with potentially the greatest impact on sea level rise are Greenland and the Antarctica. The melting of the ice sheet of Greenland alone could raise the sea level by 7 meters. Measurements in Greenland show that the ice closest to the sea is already melting on the surface, creating meltwater ponds and water streams towards the open sea. Both ice sheets are decreasing, and if temperatures continue to rise, it could only be a matter of decades before the ice sheets of Greenland are melted completely and the Antarctic ice sheets become unstable. Sea level rise is not the only outcome of melting ice (Archer & Rahmstorf, 2010). Another consequence is the loss of surfaces that reflect sunlight back into space, decreasing the earth's surface albedo and adding more heat to the earth's surface

Compared to ice sheets, mountain glaciers and ice caps contain significantly less water, but this water melts much quicker under increasing temperatures. Glaciers have been retreating since the 18th century, but only since the 1970s has the rate of the melting increased. Mountain glaciers and snow packs store winter precipitation and release it slowly over the summer, providing a fresh water source when it is needed for agricultural irrigation. With glaciers retreating and snow packs declining, fresh water from mountain streams could cause decreases in dowstream storage of water supplies. The melting of permafrost soils is another problem that has the effect of increasing the concentration of GHGs in the atmosphere. Arctic permafrost underlies almost one fifth of the planet's land surface and usually contains methane hydrate. As long as it is frozen, methane hydrate does not present any danger for the environment, but when the ice thaws due to climate change, the methane converts to a very potent heat trapping gas.

Precipitation and drought

In addition to increasing global temperature, climate change also impacts precipitation patterns. Unlike temperature, which has increased almost everywhere on the planet, precipitation is increasing in some parts of the world and decreasing in others. The warmer the air becomes, the more water it can store and then release it during colder days. This can lead to storm floods and heavy damage in areas where the infrastructure is not able to handle the release of large amounts of water in short amounts of time. The map in Figure 1.3 shows the observed changes in annual precipitation over land.

For example, annual precipitation as well as heavy rainfall increased on the east coast of the United States, but decreased in Africa, where food shortages and hunger are already a major concern. In 2012, hurricane Sandy caused strong rainfalls on the east coast causing floods, heavy damages, and even human casualties. In total, hurricane Sandy caused 117 death in the United States and 69 in Canada (CNN, 2013). According to the United States Department of Commerce (2013), the total economic loss caused by the hurricane in New Jersey for travel and tourism spending alone was estimated at $950 million. In addition, the New Jersey State Government concluded that it will cost approximately $29.5 billion to repair all damages caused by Hurricane Sandy. Besides the increases in frequency and magnitude of heavy rainfall, seasonal changes in precipitation are occurring. These changes are especially important to land ecosystems and the agricultural sector. Farmers are concerned about seasonal changes of rainfall impacting their growing and harvesting seasons. Heavy rainfalls are already delaying spring planting in some areas of the United States, jeopardizing the livelihoods of farmers. The resulting flooding of the fields during the growing season causes low oxygen levels in the soil, which destroys crops and increases the likelihood of root diseases. In addition, research suggests that increasing temperatures will most likely reduce livestock production during the summer season.

Without causing any changes in the annual average rainfall, in some regions, precipitation has decreased in the summer but increased in the winter, resulting in increasing risks of flooding during the winter and drought in the summer

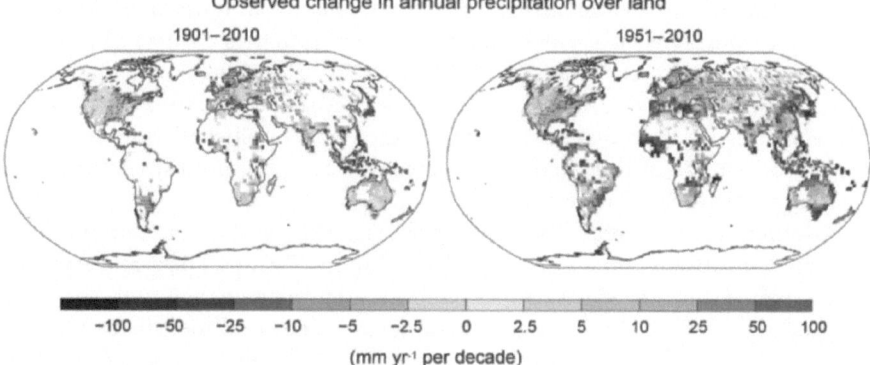

Figure 1.3 Observed change in annual precipitation over land

Source: Intergovernmental Panel on Climate Change. (2013). Summary for policymakers. In T.F., Stocker, D. Qin, G.-K. Plattner, M. Tignor, S.K. Allen, J. Boschung, . . . P.M. Midgley (eds.), *Climate change 2013: The physical science basis. Contribution of Working Group I to the Fifth Assessment Report of the Intergovernmental Panel on Climate Change* (pp. 1–30). Cambridge, UK/New York, NY: Cambridge University Press. doi:10.1017/CBO9781107415324.004.

with potentially devastating results for the agricultural sector. Seasonal changes may also impact areas that rely heavily on tourism and winter sports. In some areas, precipitation that used to fall as snow during the winter is now falling as rain. Consequently, the reduction in the snowpack not only shortens the winter sports season but also reduces the water runoff during the summer when water is most needed for agriculture. Droughts are another major result of changes in precipitation. There are different ways to define and measure the severity of droughts. The most common measurement used is the Palmer Drought Severity Index (PDSI), which considers not only the monthly amount of precipitation, but also the regional average temperatures. According to the PDSI, droughts became more severe between 1990 and 2002 and are increasing. In this index, the severity of the drought is shown in terms of minus numbers, and excess rain is reflected by plus numbers. Although heavy rainfall has increased, the risk of droughts has only decreased in very few regions of the world. Instead, the amount of dry areas has more than doubled in size since the 1970s (IPCC, 2007).

Human health, food insecurity, and ecological refugees

In addition to environmental impacts, climate change can also cause or intensify health and social issues. The impact of climate change on human health is a relatively new research field, and at this point, not much data are available. Nevertheless, existing research shows a strong correlation between heat waves and increased mortality rates. In 2003, the European heat wave was responsible for at least 35,000 deaths, many of them in highly industrialized countries that were considered less vulnerable to weather extremes compared to developing countries in Africa or South America. Furthermore, data indicate that ticks spreading Lyme disease, the Anopheles mosquito carrying Malaria, and other viruses are spreading northward. An increase of pollen allergies is another impact of climate change, since the increase in temperature causes the pollen season to start earlier in the year.

Another problem area reinforced by climate change that is becoming more and more visible is food insecurity. The United Nations Food and Agriculture Organization (FAO) warned in 2008 that climate change will negatively impact all aspects or dimensions of food security. Those dimensions consist of food availability, food accessibility, food utilization, and food system stability (FAO, 2008). Although food systems in all countries will be further impacted by climate change in the future, the world's poorest and most food insecure countries will be affected the greatest. Another measurement to assess food insecurity is the Global Food Security Index (Global Food Security Index, 2014). The main categories from which the index is compiled include (a) affordability, (b) availability, and (c) quality and safety. According to the latest ranking, the majority of the 109 countries included in this measurement have improved their food security from 2013 to 2014. However, the rankings also show that several developing nations still struggle with providing sufficient infrastructure, political unsettlement, and food inflation, which are significant barriers to reaching a satisfying level of food security. In the latest ranking from 2014, the top 10 nations are all located in North America, Asia, and Europe with the United States being on top followed by Austria, Netherlands, Norway, Singapore, Switzerland, Ireland, Canada, Germany, and France. On the other hand, the 10 countries with the lowest scores are all located in Africa such as Burkina Faso, Mozambique, Niger, Haiti, Tanzania, Burundi, Togo, Madagascar, Chad, and in last place the Democratic Republic of Congo.

According to the IPCC, as well as pointed out by UN Secretary General Ban Ki Moon (United Nations General Assembly, 2009), well-documented climate change impacts, such as severe drought conditions, increasing sea level rise, storm floods, and other extreme events, are threat multipliers that aggravate stressors abroad, including food security, poverty, environmental degradation, political instability, and social tensions. In turn, such conditions increase the likelihood of a significant uptake of environmental refugees seeking shelter in the United States or even increased terrorist activity and other forms of violence in other regions of the world as stated in the latest reports by the United Nations Security Council (UNSC, 2011), the CNA Military Advisory Board (CNA Military Advisory Board, 2014), or the Department of Defense (DOD, 2014). Moreover, President Obama's national security strategy released in February 2015 now also acknowledges climate change as a threat to national security, potentially increasing the amount of climate migration to the United States.

There is an increasing body of literature (Barnett & Adger, 2007; Bettini, 2013; Oels, 2012) that supports the hypothesis that future climate change impacts will result in an increase of climate refuges, including cross-border migration, which can lead to far-reaching domino effects that risk destabilizing whole regions (Stern, 2008). This puts unforeseeable pressure on communities, especially in border regions, which already are impacted by legal and illegal immigration in multiple ways. Additional increases in border crossings can exceed the adaptive capacity of these municipalities, posing risks to local economies, health systems, infrastructure, and other sectors, which in turn can lead to social unrest and potential threats to national security. Considering the expected increase in extreme weather events, it is therefore imperative to build up the capacity for disaster risk reduction, disaster preparedness, and conflict prevention.

Future impacts of climate change

Climate models and scenarios

Future climate change is already built into the system by past GHG emissions, which will take decades to disappear from our atmosphere. Thus, impacts will occur even if we act today and reduce GHG emissions. In fact, the first Kyoto Protocol had as a goal to reduce GHG emission by 5.2 percent below the emission levels of 1990 by 2012. This goal failed; the amount of worldwide GHG emissions is still increasing. Research calls for the increased likelihood that extreme weather events and sea level rise will increase and droughts will become longer and more severe. Additional impacts in the future might be major alterations in oceans, ice, or storms, as well as massive dislocations of species, pest outbreaks, and major shifts in wealth, technology, and societal priorities (Stern, 2008). There is a wide range of climate models available, ranging from relatively basic models that focus on the aspects of the earth's heat balance to very complex and detailed simulations, which aim to show possible future impacts of climate change under a variety of assumptions.

The models compute different outcomes or scenarios based on different assumptions regarding possible future amounts of GHG emissions, policy selections, behavioral actions, and other aspects that might impact future climate trends. It is important to understand that these models do not predict the future. They simply offer possible future scenarios. The future, to a large degree, depends on human behavior, which is impossible to predict. The scenarios, however, do provide important data to decision makers, which allow them to make better informed, long-term decisions that impact the future. Based on different scenarios, Figure 1.4 shows impacts of climate change by sector as temperatures increase. Although there are short-term benefits in some areas, in the long run, as temperatures keep increasing, the scenarios indicate significant negative consequences in all sectors.

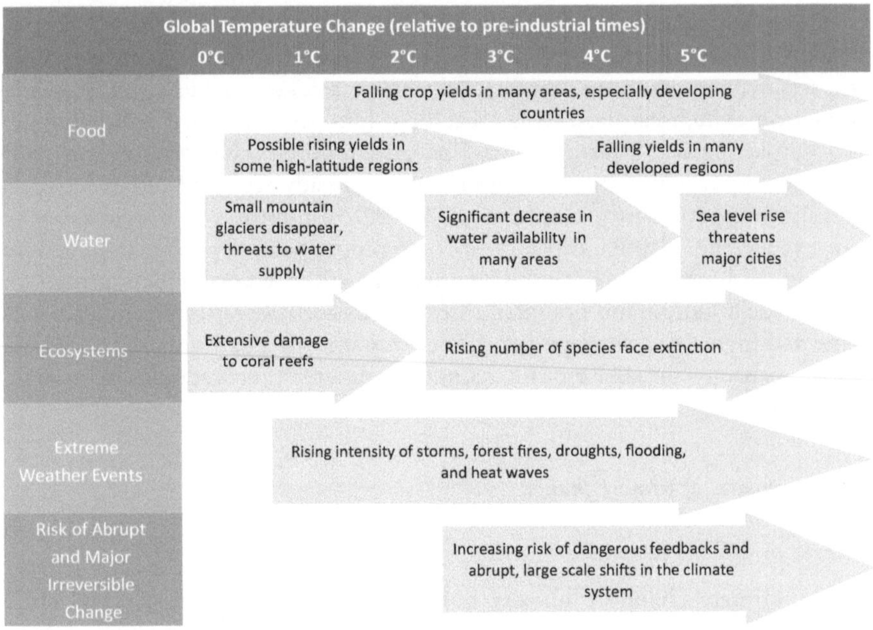

Global Temperature Change (relative to pre-industrial times)					
0°C	1°C	2°C	3°C	4°C	5°C

Food: Falling crop yields in many areas, especially developing countries

Possible rising yields in some high-latitude regions

Falling yields in many developed regions

Water: Small mountain glaciers disappear, threats to water supply

Significant decrease in water availability in many areas

Sea level rise threatens major cities

Ecosystems: Extensive damage to coral reefs

Rising number of species face extinction

Extreme Weather Events: Rising intensity of storms, forest fires, droughts, flooding, and heat waves

Risk of Abrupt and Major Irreversible Change: Increasing risk of dangerous feedbacks and abrupt, large scale shifts in the climate system

Figure 1.4 Projected impacts of climate change according to specific rises in temperature

Source: Stern, N. (2008). *The Economics of Climate Change: The Stern Review*. Cambridge, UK: Cambridge University Press.

Future global temperatures

According to the fifth IPCC Assessment Report (2013), between 1986 and 2005, the global mean surface temperature will likely increase by 0.3°C to 0.7°C between the years 2016 and 2035 compared to temperatures measured between 1986 and 2005. As shown in Figure 1.5, projections reaching further into the future suggest average temperature increases ranging from 0.3°C to 1.7°C all the way to 2.6°C to 4.8°C by the year 2100 depending on the emission scenario. Moreover, the IPCC concludes with a very high degree of certainty that there will be more hot weather extremes in the future and less cold temperature anomalies.

Future precipitation patterns

In addition to rising temperatures, changes in precipitation will also impact society and the natural environment. A secure water supply is fundamental for our food supply and for the livelihood of plants and animals. Yet, a possible future change in precipitation is much harder to predict than other features of climate change. Current climate models operate on a large spatial scale, making it very difficult to capture important regional differences in rainfall. Therefore, the uncertainties regarding possible future trends of rainfall extremes or droughts can be quite large.

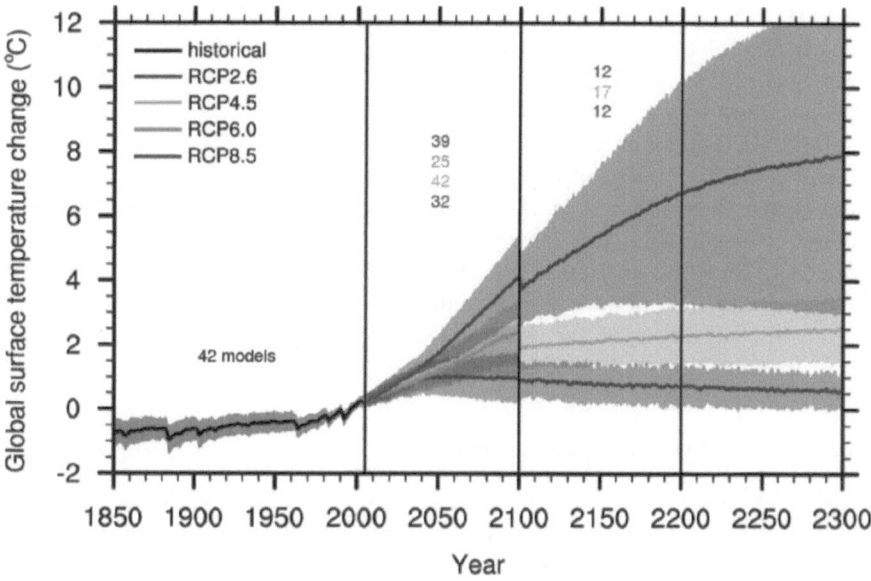

Figure 1.5 Global surface temperature change

Source: Rhein, M., Rintoul, S.R., Aoki, S., Campos, E., Chambers, D., Feely, R.A., . . . Wang, F. (2013). Observations: Ocean. In Stocker, T.F., D. Qin, G.-K. Plattner, M. Tignor, S.K. Allen, J. Boschung, . . . P.M. Midgley (eds.), *Climate change 2013: The physical science basis. Contribution of Working Group I to the Fifth Assessment Report of the Intergovernmental Panel on Climate Change* (pp. 255–316). Cambridge, UK/New York, NY: Cambridge University Press. doi:10.1017/CBO9781107415324.010.

Despite the uncertainty in terms of the severity of the precipitation changes, all models anticipate future droughts, heavy rainfalls, and floods (IPCC, 2013). The different scenarios that indicate the difference in precipitation between wet and dry regions will increase, meaning that wet areas will become wetter and dry areas will become even dryer. Furthermore, as mean surface temperatures increase, extreme precipitation events will most likely have also intensified by the end of the 21st century. Climate models show that heavy rainfall will become strong and more frequent, especially over most of the mid-latitude land masses and over wet tropical regions. Monsoon seasons are also likely to start earlier in the future and last longer in many regions. Risks of droughts on the other hand are forecasted to amplify in Australia, the eastern parts of New Zealand, as well as in the Mediterranean, central Europe, and Central America. In terms of snowfall, decreases in the length of the snow season can be expected in most of Europe and North America.

Future impacts on water security

One of the most significant impacts of climate change on human society in the future revolves around water security (IPCC, 2007). Regions in the Mediterranean, southern Africa, Western Australia, and in the southwest United States will likely face serious future droughts. According to the 2007 IPCC report, by the year 2050, 1 to 2 billion people could suffer from droughts and decreasing water quality. For example, in North America projections suggest that stream-flow in the southwest of the United States could decrease to a point that by the year 2020 present water demand cannot be satisfied anymore. In parts of South America, especially in Brazil, climate change could lead to a decrease of ground-water recharge by more than 70 percent by the 2050s. In Europe, electricity production potential at existing hydropower stations are likely to decrease by more than 25 percent by the 2070s. Overall, water stress in regards to quality and availability will increase, affecting up to two-thirds of the global land area. In turn, this will affect food security and water quality and can adversely impact human health.

Uncertainty in climate change projections

All future climate change impact scenarios are characterized by uncertainties. The uncertainty arises from the very complex climate science, possible future behaviors, decisions by humans, and from internal processes in the climate system. Future human behavior is very unpredictable and is influenced by atti-tudes towards quality of life and wealth. Future emission trends will depend heavily on the development and availability of new technologies, the imple-mentation of different environmental policies, and by their level of acceptance and support by the public. Currently, the US federal policy through executive order is advancing regulations to reduce CO_2 emissions in coal plants. Internal processes in the physical climate system might involve changes in vegetation, variations in the earth's orbit around the sun, or volcanic eruptions. Given these large uncertainties, it is very difficult to choose the appropriate adaptation strat-egies to ameliorate future climate change impacts. Traditional approaches such as making decisions based on worst case scenarios do not apply to the highly complex and uncertain issue of climate change. Instead, a more flexible frame-work is required that allows decision makers to develop strategies based on many different possible scenarios with feedback loops. This approach is referred to as advanced scenario planning and is a key component of the anticipatory governance framework (Quay, 2010).

The concept of "anticipatory governance" can be described as "a system of institutions, rules, and norms that provide a way to use foresight for the purpose of reducing risk and to increase capacity to respond to events at early rather than later stages of their development" (Fuerth, 2009, p. 29). It presents a new model for decision making, deals with high uncertainties, and consists of the anticipatory future steps and feedback creation for flexible adaptation strategies, monitoring, and action. Anticipation and future analysis are based on advanced scenario planning and include methods such as aggregated averages, risk assess-ments, sensitivity analysis of factors or decisions driving the scenarios, identifi-cation of unacceptable or worst case outcomes, and assessment of common and different impacts among the scenarios. Due to the uncertainties surrounding climate change and the changing impacts over time, the final step "monitoring and action" demands that policy makers and decision makers revise adaptation strategies on a regular basis.

From mitigation to adaptation: the shift in consideration of adaptation as one of the main response strategies of climate policy

A great amount of political intervention, public behavioral change, and support for climate action planning will be necessary in the next decade to mitigate the causes of global warming. Comprehensive changes in numerous aspects of society and the built-up environment will be required to reduce the effects of climate change that are already unavoidable. These actions collectively known as adaptation and mitigation of climate change, presents challenges to local decision makers and urban planners. Mitigation and adaptation strategies are considered the two main policy responses to climate change. However, the two are not independent, and in fact, mitigation and adaptation are driven by the same set of problems (Ausubel, 1991; Frankhauser, 1996; Smit & Wandel, 2006), and the more that mitigation takes place, the less adaptation will be needed (Huq & Grubb, 2004).

Until the year 1992, the term *adaption* was hardly used in the context of climate change or environmental risks in general (Schipper & Burton, 2009). Instead, adaptation was mainly employed in the field of Darwinian Theory as a scientific concept (Burton, 1994). Furthermore, social scientists tried to avoid using the term in the past, because adaptation tended to acquire controversial or negative associations in a social context. As an alternative for adaptation, scientists used the term *human adjustment* (White, 1945) at first to describe strategies to cope with environmental risks, such as flood control engineering, land-use regulations, building codes, watershed management, flood forecasting, warnings and evacuation, and relocation. Following human adjustment, other terminologies were introduced by scientists, for instance *coping, risk management, vulnerability reduction*, and *resilience*. Today, these concepts play an important role when considering adaptation to climate. The Intergovernmental Negotiation Committee, which worked on the draft of the United Nations Framework Convention on Climate Change (UNFCC) in 1992 returned to using the term adaptation in the context of climate change. Later that year, the draft was agreed on during the UN Conference on Environment and Development in Rio de Janeiro, Brazil. Although the term *adaption* was used several times throughout the document, there was no clear definition of the concept and more attention was paid to mitigation. The main goal stated in the UNFCC was to stabilize the GHG emissions of Annex 1 countries at 1990 levels by the year 2000 (UN, 1992).

There are several reasons why mitigation historically received much greater attention than adaptation (Fuessel, 2007). One of the main reasons is that mitigation methods reduce climate change impacts on all climate sensitive systems.

Adaptation on the other hand is limited for many systems. Another advantage from a policy perspective is that benefits are certain, because mitigation addresses the core cause of human-induced climate change: the large amount of GHG emissions. The effectiveness of adaptation strategies are influenced by high uncertainties, because they depend on the accuracy of regional climate and impact projections. In addition, GHG emissions are easy to assess and can be monitored quantitatively compared to measuring the effectiveness of adaptation, which focuses on avoiding negative impacts caused by climate change. As a result, a coherent regime for mitigation has been created. The IPCC Third Assessment Report (IPPC, 2001) emphasizes the need for stabilization of GHG concentrations, but does not present any goals for adaptation. Furthermore, the concept of mitigation is clearly understood, whereas a coherent theory and definition of adaptation is still missing today (Schipper & Burton, 2009). Moreover, many governments agreed on GHG emission targets and time frames to reach that goal. On the contrary, there are no targets and schedules for adaptation. Most importantly, compared to adaptation, mitigation has a legal instrument in the form of the Kyoto Protocol, which was initially adopted in December 1997 in Kyoto, Japan.

However, new observations and scientific projections resulted in considering adaptation as an equal if not even more important response strategy to climate change. On many fronts climate change and its impacts are occurring faster than expected (Pittock, 2009). Even if current mitigation measurements prove to be successful in reducing GHG emissions, climate change is already occurring (Raper et al., 1996; White & Etkin, 1997; Wigley, 1999). Moreover, further climate change is already built into the system by past GHG emissions. For example, it is very likely that extreme weather events and sea level rise will increase and droughts become longer and more severe. Additional impacts in the future might be major alterations in oceans, ice, or storms, as well as massive dislocations of species or pest outbreaks, or even major shift in wealth, technology, and societal priorities (IPCC, 2013). This development makes the design and implementation of planned adaptation strategies a necessity to deal with the impacts that cannot be prevented anymore and to prepare for possible future threats. In the context of climate change adaptation practices are adjustments with the aim to increase resilience or decrease vulnerability to current or expected impacts of climate change (IPCC, 2007). Without the successful implementation of adaptation strategies, it will be impossible to minimize the economic costs of climate change impacts, protect human health and welfare, and limit harm to infrastructure, ecosystems, and biodiversity (Yohe & Strzepek, 2007). Adaption measurements are required to either protect settlements threatened by sea level rise, floods, and droughts, or to enable the population to relocate. In the case of water shortage, development of new supplies or the implementation of conservation measures will become necessary. Adaptation measures also play an important role in the agricultural sector. Farmers and ranchers might need to react to the impacts of climate change by changing crops, raising different livestock, or by relocating. Furthermore, vulnerable settlements need protection from increasing heat–related events. Already occurring shifts in disease and insect pattern also emphasize the need for adaptation measures for the public health sector (Smith et al., 2010).

There are other important aspects and reasons why adaptation is now being more considered by policy makers and scientists. As discussed before, GHG emission reductions take time, and benefits will not be fully experienced for decades (Casassa & Rosenzweig, 2007). Adaptation measurements on the other hand have a much shorter lead time. Due to the strong links with development initiatives and the implementation mostly on the local or regional scale, adaptation efforts and their results are much faster visible compared to mitigation measures (Schipper, 2006). Furthermore, the efficiency of adaptation strategies is less dependent on the actions of others and does not need international agreements. The ability to implement adaptation methods on the local and regional level is highly important. The majority of the world population lives in urban environments (Bolund & Hunhammar, 1999), which have varying local circumstances and have to be addressed differently. Since 2008, more than half of the world's population live in urbanites (World Health Organization, 2008). It is projected that the number of city dwellers will rise to 5 billion people by the year 2030, with the majority of growth taking place in Africa and Asia (United Nations Population Fund, 2007). Most of this growth is likely to happen in smaller towns and cities which only have limited resources to respond to the possible negative impacts of climate change. Without proactive adaptation, these places will become even more vulnerable to future climate change.

The ability to identify probable adaptation is very important for impact and vulnerability assessment and therefore key for calculating the costs as well as the risks of climate change (Yohe et al., 1996; Tol et al., 1998; UNEP, 1998; Smit et al., 1999). Thus, it is necessary to gain a good understanding of the process of adaptation, the public's attitude and perception towards climate change, and the circumstances under which various types of adaption are likely to happen. This knowledge is essential for making informed decisions concerning the vulnerabilities of sectors, regions, and communities. Researchers and policy makers have realized that reaching the limits of adaptability creates an unacceptable or damaging situation that, at least at the local level, could be considered "dangerous" and could lead to dire economic consequences (Stern, 2008). This can only be avoided by reducing the level of climate change to stay within the limits of adaptability. Therefore, the ability to adapt is what must determine the targets for reducing GHG emissions. For this reason, understanding adaptability is vital, not only so that people can adapt where possible, but also to determine how urgently, and by how much GHG emissions must be reduced (Burton, 2004).

More recently, the awareness among policy makers and scientists grew that the past approach of looking at adaptation and mitigation as fundamentally different approaches to the same problem was wrong (Fuessel, 2007). In fact, the two main responses to climate change are complementary rather than mutually exclusive alternatives. Now, possible synergies and tradeoffs between mitigation and adaption are increasingly acknowledged (Klein & Huq, 2007). This trend of increasing political interest in integrating the two options is emphasized by the inclusion of an extra chapter in the IPCC Fourth Assessment Report (IPCC, 2007) focusing on the interaction between adaptation and mitigation. The chapter points out that these two response options can sometimes be mutually reinforcing but they can also work against each other. Overall, the last three IPCC reports play a key role in raising awareness and advocating adaptation as a suitable response strategy (IPCC, 2001; IPCC, 2007; IPCC 2013). "Climate Change 2001: Impacts, Adaptation, and Vulnerability" was published in 2001 and presented an ample overview of the possible impacts of adaptation measurements to climate change. This publication marks not only the first time that the need for adaptation was recognized by the IPCC, but also functions as an introduction to adaptation strategies on which the fourth and fifth Assessment Report build upon with chapters dedicated to adaption research and strategies.

Duality of urbanities: contributors and victim of climate extremes

Today, more people live in city environments than in rural areas. Projections suggest that by 2050, approximately 75% of the world's population will live in cities. The ongoing rapid urbanization is transforming natural landscapes and ecosystems into artificial urban landscapes, thus creating unique and complex social, economic, and environmental challenges, emphasizing the need for sustainable development patterns. The move of populations into cities is coupled with transformative changes including the growth of megacities, expansion of urban poverty, especially in developing countries, and growing concern over food security, among other issues.

Urbanities are also both contributors and impacted by climate extreme which impact climate change decision and policy processes. Urban centers are a significant net contributors to GHG's attributable to climate changes through the built environment (e.g. housing, roads, and parking lots), transport, consumption and recreation, a situation that looks to be exacerbated if current global urbanization trends continue. At the same time, frequent occurrence of extreme weather events in urban areas (e.g. torrential rainfall in London, floods in Boston) have shed stark light on the vulnerability of urban dwellers to the effects of climate change. This duality – urbanites as contributors to and impacted by extreme climatic patterns induced by climate change – sets them apart from their counterpart in rural settings. Policy makers and designers of built environments, working to improve the condition of the urban population, now face new challenges. Since 2008, more than half of the world's population has lived in an urban environment (World Health Organization, 2008), and the majority of the world's energy consumption has either occurred in cities or as a direct result of the way cities function. Cities consume about 75 percent of the world's energy and are responsible for more than 71 percent of energy-related GHG emissions (International Energy Agency, 2008). However, in regards to per capita emissions, research shows that urban dwellers usually have a smaller carbon footprint compared to residents in more rural areas (Norman et al.,

2006; Dodman, 2009).This underscores the impact of urban form and public behavior on emissions.

Today, city dwellers in urbanized areas worldwide are confronted with extreme weather events induced by the current level of GHG in the atmosphere and the resulting climate change. In many cases, people can only respond to these events by implementing measures that also produce GHG emissions. As a result, by reacting to extreme weather events, the urban population produces even more GHG emissions, expedites climate change, and intensifies future weather extremes. Such processes are often referred to in the literature as positive feedback loops, climate feedback loops, or feedback cycles (Backer et al., 2002; Karl et al., 2009; Kamal-Chaoui & Roberts, 2009). Feedback loops pose a significant concern to climate change mitigation and adaption and occur "when the increasing heat caused by elevated levels of CO_2 begins to affect certain previously undisturbed parts of the environment and the result of the disturbance is a new, distinct and additional contribution to making heat grow" (Albertson, 2009, p. 132). An example at the global scale is Arctic permafrost, which underlies almost one-fifth of the planet's land surface and usually contains methane hydrate. As long as it is frozen, methane hydrate does not present any danger for the environment, but when the ice thaws due to climate change, the methane converts to gas as another GHG. Moreover, methane hydrate is a much more effective heat-trapping gas than CO_2. Thus, the melting permafrost also contributes to climate change and growing heat and in turn causes even more permafrost to melt. Another positive feedback loop accelerating global warming involves water vapor. Due to already increasing temperatures, more water evaporates, increasing cloud cover which in turn prevents the earth surface to reflect the sun's heat back out of the atmosphere. Consequently, rising temperatures increase water vapor, allow the atmosphere to hold more moisture, and increase cloud cover, which eventually will lead to heavy evaporation and storm floods (Santer et al., 2007)

A very good example for the duality whereby urbanites are both contributors and impacted by climate extreme is the Urban Heat Island (UHI) effect (Stone Jr., 2006; Stewart & Oke, 2009; Stone Jr. & Rodgers, 2010). The UHI effect describes the phenomena of higher temperatures of both the atmosphere and surfaces in cities and urban areas compared to their non-urbanized surroundings (Jusuf et al., 2007). Heat islands are directly related to urbanization. Buildings, roads, and paved surfaces store the heat during the day and then release it slowly during the evening. As a result, the urban land is hotter than its surrounding area.

As illustrated in Figure 1.6, UHIs increase local temperatures and create a greater demand for air-conditioning, which intensifies the UHI through anthropogenic heat and increasing regional energy demand. The enhanced energy demand increases GHG emissions, which in turn further elevates near surface temperatures and requires additional energy demand for cooling. In addition to increasing climate change, UHIs also lower air quality, deteriorating public health. Overall, such feedback loops as the once discussed above are self-reinforcing cycles, which will eventually spiral out of control and further increase severe extreme weather events impacting city dwellers. Therefore, it is

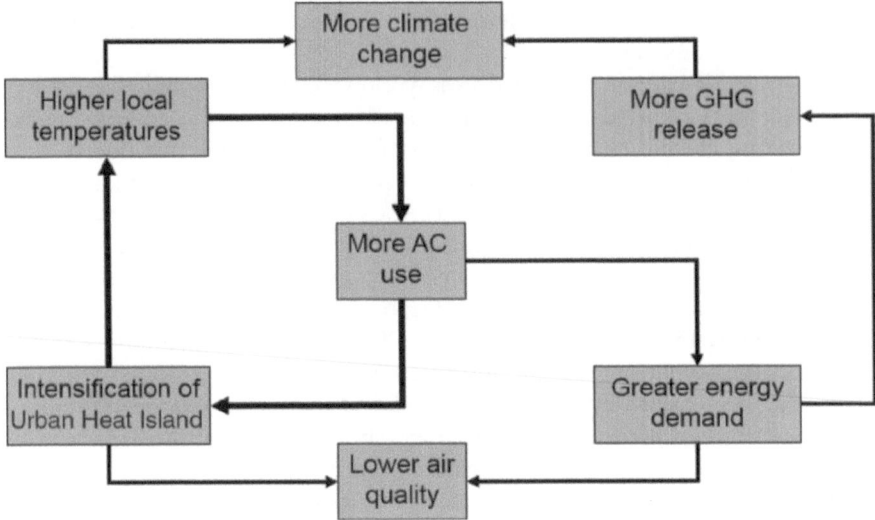

Figure 1.6 Feedback loop of the Urban Heat Island Effect

Source: Adapted from Organisation for Economic Co-operation and Development (2008).

vital that planning and decision-making processes acknowledge these issues and address them in a comprehensive manner and with a sense of urgency.

Decision makers need to implement proactive measures in forms of emission regulations and educational programs and establish mandatory density and energy efficiency criteria. In addition, planning processes have to limit the likelihood of climatic feedback loops and decrease the need for city dwellers to respond to weather extremes in ways that increase GHG emissions even further. Moreover, it is important to realize that reducing the levels of GHGs in the atmosphere to levels which would decrease current extreme weather events will take quite a long time (Pittock, 2009). Furthermore, the effectiveness of current mitigation measures is still widely discussed among scientists and associated with large uncertainties (NRC, 2009). Therefore, planning processes also have to consider the implementation of adaptation strategies to cope with weather impacts that cannot be avoided, for example, problems in coastal areas connected with the irreversible sea level rise that already occurred. This includes large investments in coastal barriers, flood protections, or robust infrastructure to manage severe storms and extreme precipitation (Bulkeley & Betsill, 2003).

In the case of mitigating the Urban Heat Island Effect, policy makers need to enforce specific building and planning practices (Figure 1.7). The existing body of research (Stone Jr. & Rodgers, 2001) points to three main policies that cities can mainstream into their planning processes without increasing GHG emissions or creating climate feedback loops. First, the increase of surface albedo within the urban environment by requiring highly reflective paving and building materials. Second, existing natural vegetation needs to be protected and

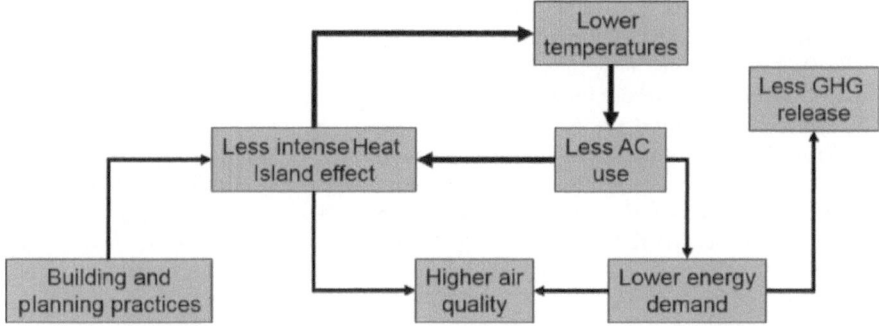

Figure 1.7 Mitigating the Urban Heat Island Effect and greenhouse gas emissions

Source: Adapted from Organisation for Economic Co-operation and Development (2008).

when possible enhanced, increasing the overall vegetation cover within the city region. Planting trees around buildings to shade urban surfaces and increasing the number of green roofs within the city limits leads to significant reduction of energy consumption for air conditioning and thus results in less GHG emissions. In addition, more vegetation also provides more sinks for carbon dioxide. Third, future developments need to be transit oriented, support walkability, and require an energy efficient building design to decrease waste heat emissions and reduce GHG emissions.

Bibliography

Albertson, R. (2009). *The Sky Is the Limit: A Brief and Easy Explanation of Climate Change for Present and Future Voters*. London, UK: The Circle Works.

Archer, D., & Rahmstorf, S. (2010). *The Climate Crisis: An Introductory Guide to Climate Change*. Cambridge, NY: Cambridge University Press.

Ausubel, J.H. (1991). Does climate still matter? *Nature, 350*(6320), 649–652.

Backer, L.A., Brazel, A.J., Selover, N., Martin, C., McIntyre, N., Steiner, F.R., ... Musacchio, L. (2002). Urbanization and warning of Phoenix (Arizona, USA): Impacts, feedbacks and mitigation. *Urban Ecosystems, 6*, 183–203.

Barnett, J., & Adger, W.N. (2007). Climate change, human security and violent conflict. *Political Geography, 25*, 639–655.

Bettini, G. (2013). Climate barbarians at the gate? A critique of apocalyptic narratives on "climate refugees". *Geoforum, 45*, 63–72.

Bolund, P., & Hunhammar, S. (1999). Ecosystem services in urban areas. *Ecological Economics, 29*(2), 293–301.

Bulkeley, H., & Betsill, M. (Eds.). (2003). *Cities and Climate Change: Urban Sustainability and Global Environmental Governance*. London, UK: Routledge.

Burton, I. (1994). Deconstructing adaptation . . . and reconstructing. *Delta, 5*(1), 14–15.

Burton, I. (2004). Climate Change and the Adaptation Deficit. In E.L.F. Schipper & I. Burton (Eds.), *The Earthscan Reader on Adaptation to Climate Change* (pp. 1–8). London, UK: Earthscan.

Calthorpe, P. (2010). *Urbanism in the Age of Climate Change*. Washington, DC: Island Press.

Casassa, G., & Rosenzweig, C. (2007). Assessment of observed changes and responses in natural and managed systems. In Intergovernmental Panel on Climate Change (Ed.), *Climate Change 2007: Climate Change Impacts, Adaptation, and Vulnerability* (Chapter 1). Cambridge, UK: Cambridge University Press.

CNA Military Advisory Board (CNA). (2014). *National Security and the Accelerating Risks of Climate Change*. Alexandria, VA: CNA Corporation.

CNN. (2013). Hurricane Sandy Fast Facts. http://www.cnn.com/2013/07/13/world/americas/hurricane-sandy-fast-facts/ (accessed 6/5/2014)

Department of Defense (DOD). (2014). *2014 Climate Change Adaptation Roadmap*. Alexandria, VA: Office of the Deputy Under Secretary of Defense for Installations and Environment.

Dodman, T. (2009). Blaming cities for climate change? An analysis of urban greenhouse gas emissions inventories. *Environment and Urbanization, 21*(1), 185–201.

Food and Agriculture Organization (FAO). (2008). *Climate Change and Food Security: A Framework Document*. Rome, Italy: Food and Agriculture Organization of the United Nations.

Frankhauser, S. (1996). The potential costs of climate change adaptation. In J.B. Smith, N. Bhatti, G.V. Menzhulin, R. Benioff, M. Campos, B. Jallow, ... & R.K. Dixon (Eds.). *Adapting to Climate change: An International Perspective* (pp. 80–96). New York, NY: Springer Verlag.

Fuerth, L.S. (2009). Foresight and anticipatory governance. *Foresight, 11*(4), 14–32.

Fuessel, H.M. (2007). Adaptation planning for climate change: Concepts, assessment approaches, and key lessons. *Sustainability Science, 2*, 265–275.

Global Food Security Index. (2014). *Ranking and Trends. New York: The Economist Intelligence Unit.* http://foodsecurityindex.eiu.com/Index (accessed 6/5/2014)

Huq, S., & Grubb, M. (2004). *Scientific Assessment of the Interrelationships of Mitigation and Adaptation*. IPCC Expert Meeting. Geneva, Switzerland.

International Energy Agency (IEA). (2008). *World Energy Outlook 2008*. Paris, France: Organisation for Economic Co-operation and Development/IEA.

International Transport Forum. (2010). *Reducing Transport Greenhouse Gas Emissions: Trends & Data*. Leipzig, Germany: The Organization for Economic Co-operation and Development.

IPCC. (2001). *Climate Change 2001: Third Assessment Report of the Intergovernmental Panel on Climate Change*. Cambridge, UK: Cambridge University Press.

IPCC. (2007). *Climate Change 2007: The Physical Science Basis. Contribution of Working Group 1 to the Fourth Assessment Report of the Intergovernmental Panel on Climate Change*. Cambridge, UK/New York, NY: Cambridge University Press.

IPCC. (2013): *Climate Change 2013: The Physical Science Basis. Contribution of Working Group 1 to the Fifth Assessment Report of the Intergovernmental Panel on Climate Change*. Cambridge, UK/New York, NY: Cambridge University Press.

IPCC. (2014a). *Climate Change 2014: Impacts, Adaptation, and Vulnerability. Contribution of the Working Group 2 to the Fifth Assessment Report of the Intergovernmental Panel on Climate Change*. Cambridge, UK/New York, NY: Cambridge University Press.

IPCC. (2014b). *Climate Change 2014: Mitigation of Climate Change. Contribution of the Working Group 2 to the Fifth Assessment Report of the Intergovernmental Panel on Climate Change*. Cambridge, UK/New York, NY: Cambridge University Press.

Jusuf, K., Wong, S., Hagen, N.H., Anggoro, E.S., & Hong, Y. (2007). The influence of land use on the urban heat island in Singapore. *Habitat International, 31*(2), 232–242.

Kamal-Chaoui, L., & Robert, A. (Eds.). (2009). *Competitive Cities and Climate Change*. Paris, France: Organisation for Economic Co-operation and Development.

Karl, R.T., Melillo, J.M., & Peterson, T.C. (Eds.) (2009). *Global Climate Change Impact in the United States*. New York, NY: Cambridge University Press.

Klein, R.J.T., & Huq, S. (2007). Inter-relationships between adaptation and mitigation. In Intergovernmental Panel on Climate Change (Ed.), *Climate Change 2007: Climate Change Impacts, Adaptation, and Vulnerability* (Chapter 18). Cambridge, UK: Cambridge University Press.

National Aeronautics and Space Administration (NASA). (2010). *NASA Research Finds Last Decade Was Warmest on Record, 2009 One of Warmest Years*. http://www.nasa.gov/home/hqnews/2010/jan/HQ_10–017_Warmest_temps.html (accessed 5/12/2015)

National Oceanic and Atmospheric Administration (NOAA). (2005). *Stratospheric Ozone Layer Depletion and Recovery*. http://www.esrl.noaa.gov/research/themes/o3/ (accessed 6/4/2013)

NOAA. (2013). *State of the Climate: Global Analysis – Annual 2013*. http://www.ncdc.noaa.gov/sotc/global/2013/13 (accessed 5/12/2015)

National Research Council (NRC). (2009). *Driving and the Built Environment: The Effects of Compact Development on Motorized Travel, Energy Use, and CO2 Emissions*. Special Report 298: Committee for the Study in the Relationships among Development Pattern, Vehicle Miles Traveled, and Energy Consumption.

NRC. (2010). *Advancing the Science of Climate Change*. Washington, DC: The National Academic Press.

Norman, J., MacLean, H.L., & Kennedy, C.A. (2006). Comparing high and low residential density: Life-cycle analysis of energy use and greenhouse gas emissions. *Journal of Urban Planning and Development, 132*(1), 10–21.

Oels, A. (2012). From the 'securitization' of climate change to the 'climatization' of the security field: Comparing three theoretical perspectives. In Scheffran, J., Brzoska, M., Brauch, H.G., Link, M., and Schilling, J. (Eds.), *Climate Change, Human Security and Violent Conflict. Challenges for Societal Stability*. Berlin: Springer.

Organisation for Economic Co-operation and Development (OECD). (2008). *Competitive Cities and Climate Change: OECD Conference Proceedings Milan, Italy, 9–10 October 2008*. Paris, France: Author.

Pittock, A.B. (2009). *Climate Change: The Science, Impacts and Solutions*. London, UK: Earthscan.

Quay, R. (2010). Anticipatory governance: A tool for climate change adaptation. *Journal of the American Planning Association, 76*(4), 496–511.

Raper, S.C.B., Wigley, T.M.L., & Warrick, R.A. (1996). Global sea level rise: past and future. In J.D. Milliman & B.S. Haq (Eds.), *Sea Level Rise and Coastal Subsidence: Causes, Consequences and Strategies* (pp. 11–45). Dordrecht, The Netherlands: Kluwer Academic Publishers.

Santer, B.D., Mears, C., Wentz, F.J. Taylor, K.E., Gleckler, P.J., Wigley, T.M.L., . . . & Wehner, M.F. (2007). Identification of human-induced changes in atmospheric moisture content. *PNAS, 104*(39), 15248–15253.

Schipper, E.L.F. (2006). Conceptual history of adaptation to climate change under the UNFCCC. *RECIEL, 15*(1), 82–92.

Schipper, E.L.F., & Burton, I. (2009). Understanding adaptation: Origins, concepts, practice and policy. In E.L.F. Schipper & I. Burton (Eds.), *The Earthscan Reader on Adaptation to Climate Change* (pp. 1–8). London, UK: Earthscan.

Smit, B., Burton, I., Klein, R.J.T., & Wandel, J. (1999). The science of adaptation: A framework for assessment. *Mitigation and Adaptation Strategies for Global Change, 4*, 199–213.

Smit, B., & Wandel, J. (2006). Adaptation, adaptive capacity and vulnerability. *Global. Environmental Change, 16*, 282–292.

Smith, J. B, Vogel, J.M., Cruce, T.L., Seidel, S., & Holsinger, H.A. (2010). *Adapting to Climate Change: A Call for Federal Leadership*. Prepared for the PEW Center on Global Climate Change. Arlington, VA: The Rockefeller Foundation.

Stern, N. (2008). *The Economics of Climate Change: The Stern Review*. Cambridge, UK: Cambridge University Press.

Stewart, I., & Oke, T. (2009). *Newly Developed "Thermal Climate Zones" for Defining and Measuring*. http://ams.confex.com/ams/pdfpapers/150476.pdf (accessed 5/07/2015)

Stone, Jr., B. (2006). Physical Planning and Urban Heat Island Formation: How Cities Change Regional Climates. In M. Ruth (Ed.), *Smart Growth and Climate Change: Regional Development, Infrastructure and Adaptation* (pp. 318–341). Northhampton, MA: Edward Elgar Publishing.

Stone, Jr., B., & Rodgers, M.O. (2001). Urban form and thermal efficiency: How the design of cities influences the urban heat island effect. *Journal of the American Planning Association, 67*(2), 186–198.

Tol, R.S.J., Frankhauser, S., & Smith, J.B. (1998). The scope for adaptation to climate change: What can we learn from the impact literature? *Global Environmental Change, 8*(2), 109–123.

United Nations (UN). (1992). *United Nations Framework Convention on Climate Change*. Bonn, Germany: Author.

United Nations Environment Program (UNEP). (1998). *Handbook on Methods for Climate Impact Assessment and Adaptation Strategies*. Amsterdam, The Netherlands: Institute for Environmental Studies.

United Nations General Assembly (UNGA). (2009). *Climate Change and Its Possible Security Implications*. Report of the Secretary General. A/64/350. New York, NY: United Nations.

United Nations Population Fund (UNFPA). (2007). *State of World Population 2007, Unleashing the Potential of Urban Growth*. New York: United Nations Population Fund.

United Nations Security Council (UNSC). (2011). *Minutes 6587th Meeting, Wednesday, 20 July 2011*, New York, S/PV.6587.

United States Department of Commerce. (2013). *Economic Impact of Hurricane Sandy*. Washington, DC: Economics and Statistics Administration. http://www.esa.doc.gov/Reports/economic-impact-hurricane-sandy (accessed 5/20/2014)

US Green Building Council (USGBC). (2012). *Green Building Facts*. http://www.usgbc.org/articles/green-building-facts (accessed 6/4/2014)

White, G. F. (1945). Human Adjustment to Floods. *Department of Geography Research Paper, 29*. Chicago, IL: University of Chicago.

White, R., & Etkin, D. (1997). Climate change, extreme events and the Canadian insurance industry. *Natural Hazards, 16*, 135–163.

Wigley, T.M.L. (1999). *The Science of Climate Change: Global and U.S. Perspectives*. Washington, DC: Pew Center for Climate Change.

World Health Organization (WHO) (2008). *The World Health Report, Primary Health Care, Now More Than Ever*. Geneva, Switzerland: World Health Organization.

Yohe, G., Neumann, J., Marshall, P., & Amedon, A. (1996). The economic costs of sea-level rise on developed property in the United States. *Climate Change, 32*, 387–410.

Yohe, G., & Strzepek, K. (2007). Adaptation and mitigation as complementary tools for reducing the risk of climate impacts. *Mitigation and Adaptation Strategies for Global Change, 12*(5), 727–739.

2 Current approaches and barriers to climate change mitigation and adaptation

The role of planners and policy makers in minimizing the negative impacts of climate change

Considering the complexity, uncertainties, and scale of possible climate change impacts, there is agreement that urban planning has the capacity to facilitate the development and implementation of adaptation as well as mitigation strategies (American Planning Association [APA], 2008). The way urban environments develop will determine whether or not a low carbon, climate resilient future, and worldwide sustainable development can be achieved. The land use planning system provides the framework to reduce greenhouse gas (GHG) emissions considerably by addressing central issues such as community design, transportation networks and use, and increasing development density (DEFRA, 2005; Intergovernmental Panel on Climate Change [IPCC], 2014). Therefore, urban planning is a key component because of its comprehensive and long-term approach to the built environment. However, traditional approaches in planning are not enough to mitigate and adapt to climate change. The increasing extreme weather events caused by climate change demand the development of new strategies which improve the resiliency of cities and their inhabitants and identify new innovative opportunities for a sustainable development pattern (Biesbroek et al., 2009). Planners have to make sure that new developments and long-term infrastructure, such as commercial and residential buildings, roads and ports, or water and transport networks, are constructed to endure possible negative climate impacts and weather hazards. In addition, long-lived infrastructure also needs to be designed to decrease energy consumption and GHG emissions of the built environment (McEvoy et al., 2006; Hallegatte et al., 2008; Satterthwaite, 2008). Planning can also play an important role in impacting public behavior, thus slowing the pace of climate change and allowing the development and implementation of adaptation measurements. The highly influential Stern Review (2008) argues that planning can be an important tool for promoting private and public investments towards locations that are less vulnerable to climate risks. Moreover, planning is distinctively qualified to provide comprehensive and long-term approaches that are required to reduce these vulnerabilities through various land use and infrastructure adjustments and zoning. Although the discussion about the role of spatial and urban planning for responding to climate change has just begun, there is a growing sense that planning will be receiving increasing attention as an important policy instrument for addressing both the causes and impacts of climate change (Bulkeley, 2006; National Research Council [NRC], 2009, PROVIA, 2013).

On many fronts, climate change and its impacts are occurring faster than expected (IPCC, 2013). To stop the rate of GHG concentrations from increasing, emissions of carbon dioxide, especially from burning of fossil fuels, need to be reduced by more than 80 percent relative to present emissions. This will take several decades to achieve without disrupting human society. Even if current mitigation practices prove to be successful in reducing GHG emissions, further climate change is already built into the system by past GHG emissions. This makes the design and implementation of planned adaptation strategies a necessity to deal with impacts that cannot be prevented and to prepare for possible future threats by increasing urban resiliency (Burton, 1996; Smit et al., 1999; Pittock, 2009). Unfortunately, climate change does not receive a high priority compared to the many tangible, visible, immediate, and urgent needs countries and governments face today (Handmer, 2003). Nevertheless, climate-related disasters are consistently drawing attention to climate change forcing the issue up the political priority list. A growing sense of urgency is emerging and clarity about the choices and planning options are essential. Whether or not the Katrina and Sandy Hurricanes are the products of climate change, the fact that these two storms were so extreme, their devastation of properties so pervasive, and recovery so difficult, argues for an important role for adaptation planning.

When it comes to climate change, mitigation planning is mainly considered a tool for influencing energy demand and GHG emissions in two ways: first, through the design of new developments and urban retrofitting; and second, through policies on location and access (Bulkeley, 2006). Policies include, for example, the promotion of energy efficiency, passive solar gain, and advancing renewable energy alternatives. The built environment and urban form influences significantly the amount of energy use and thus, level of GHG emissions which are the main cause for climate change (Bulkeley & Betsill, 2003). The term *urban form* describes the spatial characteristics of fixed elements within a metropolitan region (Anderson et al., 1996). Urban form not only includes land uses and their structure and densities, but also the spatial design of transport and communication infrastructure. As pointed out by Calthorpe (2010), however, urban design is still often neglected as a strategy against climate change. Moreover, he argues that without more sustainable development, patterns mitigating and adapting to climate change will be impossible. Nevertheless, urban planning is recently receiving increasing attention as an instrument to change urban forms characterized by urban sprawl and promote a more sustainable development pattern (NRC, 2009).

Adapting to climate change is at least as important as mitigating it. As a result, the field of climate change adaption is gaining importance and increased consideration within the scientific community and policy makers (Smit et al.,

2000). The increasing consideration of adaptation strategies in climate policy is not only a global issue, but also a significant challenge at the local and regional scales. In the past two decades, climate change evolved from being strictly viewed as an environmental problem to being addressed as a human-influenced development issue (Biesbroek et al., 2009). As a result, the awareness of possible climate change impacts is now guiding the sustainable development debate, which is increasingly criticizing the current development pathways while emphasizing the need for alternative development policy responses at different scales and in different places (Davoudi et al., 2009). This ongoing discussion also impacts the work of urban and spatial planners, because there is a special need to focus on cities. The concentration of many activities, people, wealth, and locations of the largest global cities in exposed areas make cities and their infrastructure particularly vulnerable to the possible impacts of climate change. Impacts of climate change can no longer be prevented anymore, and planners will need to address effects such as rising sea levels, greater drought conditions, and flood control (Gill et al., 2007). This has recently gained greater traction as a result of the devastation caused in the Northeast United States due to Hurricane Sandy.

Due to the increasing attention to climate change adaption from scientists and politicians, urban planners are arguing to include stronger climate change considerations in the planning process (Greiving & Fleischhauer, 2008). Thus, preparing cities for the impacts of climate change is an increasing and additional challenge for urban planning. Nonetheless, planning for climate change adjustment is still in its early stages, but is gaining greater attention and resources from states and cities (PEW Center, 2008). Within the planning community, the discussion about the role of spatial and urban planning for responding to climate change has just begun (Bulkeley, 2006; Biesbroek et al., 2009). It is important to understand that there is no globally applicable strategy for climate change adaptation (Fuessel, 2007; The Heinz Center, 2007). Instead, they can vary significantly depending on geographical location (IPCC, 2013). Successful adaptation methods in one place cannot be fully transferred or generalized for other jurisdictions. Thus, adaptation is mainly a local effort and urban planners need to take local circumstances and vulnerabilities into account very carefully when developing and implementing climate adaptation strategies.

Planning for mitigation

In the past two decades, planning approaches such as New Urbanism, Traditional Neighborhood Design, Smart Growth, Sprawl Repair, or Transit Oriented Development have emerged as alternatives to conventional patterns of urban design in the United States (Ellis, 2002). Influential literature and advocators of these planning strategies (Ruth, 2006; Ewing et al., 2007; Calthorpe, 1993; Tachieva, 2010) argue for introducing design features, that when implemented will reduce energy and car use. These principles comprise the mixing of land uses, compact urban development, walkable neighborhoods, and transportation alternatives. In addition, New Urbanism, Traditional Neighborhood Design, and Transit Oriented Development promote the return to a gridded street system. Shorter trip lengths and improved accessibility are supportive of non-motorized transport. Overall, the existing literature underscores the importance of urban form and household energy use as well as underlines the increasing responsibility of planners to advocate and implement more sustainable development patterns which reduce energy uses and thus GHG emissions.

Urban form and transportation

One of the first studies addressing the link between land use and transportation and its implications for planners (Kelly, 1994) argues that decisions made in the transportation sector influence land-use patterns and vice versa. This conclusion was based on a comprehensive literature review covering six decades of research, examining case studies, practical recommendations, and mainly theoretical work (American Society of Planning Officials, 1951; Wingo, 1961; Schaeffer & Sclar 1975; De la Barra 1989). Due to the strong focus on theoretical literature and thus the lack of available data on the relationship between urban form and travel behavior, the study did not allow the validation of the benefits of the presented theories, designing ideas, and polices. Another earlier study that tested the hypothesis that a more transit-oriented development pattern will reduce the use of the car and overall automobile dependency was published by Cervero and Gorham (1995). Similar to the study by Kelly (1994), this study also was not based on quantifiable data. At the time, these studies were conducted projects, and neighborhoods that incorporated the ideas of transit oriented and compact development were either unfinished or too new. Therefore, in-depth examinations of the transportation impacts of traditional neighborhood developments were not possible at the time the articles by Kelly (1994) or Cervero and Gorham (1995) were written. Such studies require comprehensive databases, which were not available at this point in time. Instead, Cervero and Gorham compared travel behavior between neighborhoods characterized by high car dependency on the one hand and transit-oriented communities on the other. In this case, the automobile-oriented neighborhoods were typical sprawl developments, the neighborhoods characterized as transit-oriented averaged higher densities and had demonstrated more gridded street patterns. The results suggest that transit-oriented or more compact development neighborhoods reduce car use and increase walkability and public transit ridership.

Nevertheless, early studies, which were mostly theoretically based, were not able to show conclusively to what degree particular settlement structures and urban design features systematically reduce car use and daily miles travelled and thus reduce transportation-related GHG emissions. Instead, uncertainties and conflicting conclusions remained. For example a study by Cervero and Gorham (1995) focusing on work trips only, not only argues for the benefits of transit-oriented and more compact development, but also recognizes that trip generation rates tend to be higher in such neighborhoods compared to auto-oriented communities due to shorter trip lengths. On the other hand, the literature review presented by Berman (1996) addresses the importance of non-work trips for reducing automobile traffic. The author suggests that a denser development pattern can only reduce driving significantly if residents have a wide range of activities within walking distance. Compared to work trips, which are relatively fixed, non-work trips are more flexible and more likely impacted by changing conditions in zoning and urban form. The study recognizes that close to 80 percent of all trips are non-working trips, with the majority being related to shopping. Thus, planning strategies that reduce non-work trips can decrease the total amount of travel by significant amounts.

Another study emphasizing the uncertainty regarding relationships between urban form and transportation was conducted by Crane and Crepeau (1998). Acknowledging the growing popularity of smart growth and similar planning concepts as tools to reduce negative environmental impacts of urban development, the authors emphasize that the impact of any specific neighborhood feature on travel behavior is still based on unproven hypotheses. The literature review published by Crane in the year 2000 follows up on the question whether neighborhood design can improve traffic conditions (Crane, 2000). Here, the author argues that the literature has made significant progress in terms of identifying key questions to understand the complex relationship between urban form and transportation. By the year 2000, the study suggests that while the body of research is improving, strong evidence that transit- or pedestrian-oriented neighborhood plans can effectively reduce the use of the automobile is still missing. Nevertheless, Crane acknowledges that existing research (Boarnet & Crane, 2001) provides some evidence that certain street patterns along with commercial concentration results in fewer non-working automobile trips, still, the scientific foundation for accepting urban design to change travel behavior is not fully accomplished yet.

Handy (2005) adds to this mostly theoretical discourse by addressing the existing body of knowledge in terms of how well the benefits of various planning and urban design strategies are supported by scientific evidence. Handy, argues that theoretical research and casual observation implies that transportation and land use are linked to each other in at least two ways. Transportation investments and policies impact development patterns, and settlement structures influence travel behavior. He also points out, however, that significant improvements in data on travel behavior and characteristics of the built environment are needed to determine which strategies are most effective to impact travel behavior. Yet the paper presents four conclusions, which appear reasonable. For example, Handy supports the argument that new highways or a higher capacity of existing highways will influence development and growth patterns, but that new highway capacity will most likely increase traffic. However, light rail transit encourages higher densities under certain conditions including experiencing significant growth and that the light rail has a major impact to the accessibility of the locations it serves. Hardy recognizes that the proper allocation of investments also offers a strategy to mitigate sprawl and reduce energy use and resulting GHG emissions. Based on the research results reviewed, Handy concludes that planning and urban design approaches have the potential to reduce automobile use.

Not all studies are based on theoretical arguments. Instead, numerous studies have emerged in recent years which report quantitative findings in regards to the impact of urban form and travel behavior or volume of traffic. However early quantitative studies (Crane, 1996; Crane, 2000; Crane & Crepeau, 1998) did not provide any evidence that the amount of density or type of street network impacted travel behavior. Although these studies concluded that certain urban design options most likely discourage driving for some kind of trips, authors also emphasized that the aggregated effect of transit-oriented or walkable neighborhood designs remain uncertain. The authors conclude that more research is needed focusing on the impact of street patterns on car and pedestrian travel in order to prove any environmental benefits from implementing new urbanism and similar planning strategies. The case study by Handy and Clifton (2001) concludes that reducing travel distance and encouraging alternative modes of travel through local shopping does not necessarily reduce automobile dependency. Focusing on non-work travel behavior of six neighborhoods in Austin, Texas, the study addresses the question whether or not a move to local shopping opportunities reduces automobile dependency. Based on quantitative as well as qualitative data collected from a travel survey, the authors point out that even if residents walk to a store on a regular basis, the overall reduction in driving trips is still relatively small. Data suggest that local shopping trips do shorten trip lengths, but some of the walking trips are made in addition to driving rather than instead. Nevertheless, Handy and Clifton acknowledge that traditional neighborhoods have more local shopping opportunities compared to other types of neighborhoods, which gives residents the choice to drive less if they want to.

Newman and Kenworthy (2006) also point out that in terms of sustainable neighborhood designs, a gap still exists between their rhetoric and reality. The authors criticize that many new urbanist developments do not fulfill their promise of car traffic and dependency. Based on collected long-term data from numerous cities worldwide and an extensive literature review, Newman and Kenworthy quantify the required level of urban density. In order to reduce car use, redevelopments or new developments must be implemented within pedestrian catchment zone, or Ped Shed, of 300 hectares around local centers or public transportation nodes, and 3,000 hectares around town centers. Furthermore, these projects need to have at least 10,000 and 100,000 residents plus jobs. Research also points to negative relationships between density and affordable housing (Alexander & Tomalty, 2002), which also threatens new urbanism, transit-oriented development, or traditional neighborhood design to be recognized as a successful and sustainable planning strategy.

However Ewing et al.'s (2008) paper "Urban Development and Climate Change", shows significant progress in reducing uncertainty regarding the impact of land use on travel behavior. The research focuses on two important questions in terms of successfully mitigating global warming. First, how much vehicle miles traveled can be reduced by implementing compact development patterns instead of continuing urban sprawl? Second, what is the impact of less vehicle miles traveled on the total amount of GHG emissions within the transportation sector? Using four different types of studies (aggregated travel studies, disaggregated travel studies, regional simulation studies, and project simulation studies), the authors suggest that the average vehicle miles traveled can be reduced by 20–40 percent by implementing design principles and guidelines that foster more compact development. Thus, Ewing et al. suggest a 7–10 percent reduction in GHG emissions by the year 2050 through changes in urban form. The proposed reduction of GHG emissions is based on plausible yet also conservative assumptions. The authors point out that a 7–10 percent reduction does not consider future positive impacts, such as reductions resulting from intelligent transportation systems, congestion pricing, pay-as-you drive insurance, possible shifts in funding away from highways towards public transit, and other complementary urban strategies. The article also points out that a compact development pattern would not only reduce GHG emissions in the transportation sector, but would also reduce the amount of harmful emissions in other sectors of the economy. Although a 7–10 percent reduction does not seem much at first, it is worth the effort for a number of reasons. First, the authors discount improved vehicle and fuel technology alone. They argue that if the current development pattern is continued, every improvement in technology will be offset by the growth of car dependency and vehicle miles traveled.

Khattak and Rodriguez (2005) also concluded that traffic reduction can be achieved in well planned and developed neighborhoods that fallow the guidelines advocated by New Urbanism, Smart Growth, or Transit Oriented Development. The authors focused on the question of whether residents of walkable and transit-oriented neighborhoods substitute walking for driving trips or if they made more trips overall. Compared to the earlier studies by Kelly (1994) or Cervero and Gorham (1995), which lacked quantifiable data to validate their conclusions, Khattak and Rodriguez implemented a quasi-experimental research design to examine and compare travel behavior of two very different but adjacent neighborhoods in North Carolina. Despite several similarities, such as sharing the same school and transit provider, one neighborhood is characterized by a traditional neighborhood design, and the other, a conventional suburban development. Data for a two-stage regression analysis were completed from a survey of 453 households. The empirical evidence suggested differences between the more compact and walkable neighborhood and conventional suburban development. Although the results do not show significant differences in the total amount of trips made in both neighborhoods, the data indicate that the choice of transportation is different: 74.8 percent of the trips by residents of the traditional community were made by car compared to 89.9 percent in the conventional neighborhood. This suggests that residents living in the traditional neighborhood substitute alternative modes for driving trips. Another benefit is the lower amount of vehicle miles travelled. In the case of the two neighborhoods in North Carolina, residents in the suburban neighborhood traveled on average 14.7 miles more per day than their counterparts. In addition, people from the traditional neighborhood undertook 1.8 fewer external trips per day.

Another key publication on the relationship between build environment and transportation was the report "Driving and the Built Environment: The Effect of Compact Development on Motorized Travel, Energy Use, and CO_2 Emissions" by the National Research Council (NRC, 2009). The NRC study concludes that a higher residential and employment density probably results in less vehicle miles traveled. In the best case, a reduction of 25 percent can be achieved. However, the report only expects a reduction of vehicle miles traveled of about 5 percent to 12 percent. Furthermore, the paper confirms that compact mixed-use development can directly and indirectly reduce energy consumption and CO_2 emissions. The authors point out numerous obstacles that make the implementation of a more sustainable land use development pattern difficult, such as the aversion of many local governments to revise their zoning codes or the lack of power within regional governments to regulate land use. Overall, the report by the National Research Council encourages the implementation of more mixed-use development to reduce vehicle miles traveled, energy consumption, and GHG emissions. However, the report is not without criticism. Scholars such as Ewing et al. (2009) criticize the projection by the National Research Council as too conservative. Instead, he points to his own studies summarized in his book *Growing Cooler* (Ewing et al., 2007). The results in this book suggest a 20 percent to 40 percent reduction of vehicle miles travelled through compact development. Moreover, the publication identifies five key factors of urban design that reduces travel and emphasizes the important role of urban planning in decreasing energy use and GHG emissions. These guidelines are now often referred to as the five *D*s (Hamin & Gurran, 2009; NRC, 2009). The first *D* represents density, which captures the amount of population and employment per square mile or per developed acre. The second *D* is diversity and is an indicator for the mix of different land uses in an area. Design, the third *D*, addresses the features enabled by the overall neighborhood layout and street design that enhance the walkability and bicycle friendliness. The fourth *D* stands for destination accessibility. This indicator assesses the effort for the population to travel between trip origins and desired destinations. The final indicator and fifth *D* is the distance to transit, which examines the accessibility to transit.

Social aspects and public attitudes to live in traditional, more walkable, and transit-oriented neighborhoods also have to be considered when discussing the benefits of specific urban designs to reduce energy use in the transportation sector. Lund (2006) conducted a study surveying households located in communities characterized by transit-oriented development. Lund points out that households moved into these neighborhoods based on a wide range of motivations. A significant factor, however, is that one-third of the households named access to transit as one of the top reasons for living in a transit-oriented development. Other responses that were equally or even more frequent pointed to lower housing costs and the overall quality of the neighborhood. Lund argues that people's attitude and lifestyle preferences influence significantly their choice of residential location. However, the author points out that it is not clear to what extent those attitudes and preferences impact the resident's daily routine, such as travel behavior compared to the opportunities provided by their household location. Nonetheless, the results of the study show that residents of transit-oriented developments use public transit more often compared to people living elsewhere.

In 2007, Boer et al. examined different design guidelines in terms of their influence on walking (Boer et al., 2007). Among the tested guidelines are four of the New Urbanism Smart Scorecard criteria: housing density, mixed land use, street network, and block lengths. The authors claim that this study is one of the first to empirically test if certain design guidelines promote walkable communities. Using different regression models, the results show that three of the four Smart Scorecard measurements promote walkable communities. However, in terms of the relationships between housing density and the resident's willingness to walk, the study produced conflicting results. In this case, neighborhoods characterized by a housing density of at least 14 units per acre promote walking. However, the results show that when housing density is between 11 and 14 units per acre the probability to walk was lower than for densities between 7 and 11 units per acres. One reason for conflicting results is that the analysis is limited to only physical aspects. Thus, it is very likely that also social aspects such as crime influence the decision to walk. Therefore, more research is needed to fully predict the impact of design guidelines on travel behavior and mode choice.

Urban form and household energy use

Holden and Norland (2005) point to population density as well as size, age, and type of housing as key factors for energy demand. Their study shows that in regards to compact urban form the type and grouping of housing are likely to be the two most important land use characteristics energy consumption related to heating and cooling. The authors also conclude that multifamily housing is more energy efficient than single dwellings. Especially residents of larger and older buildings have a higher energy demand than their counterparts living in smaller, but also newer units which are based on the latest building designs, materials, and technologies. These findings are echoed by Ewing and Rong (2008). Furthermore, the energy required for establishing electricity transmission and distribution is higher in sprawling communities compared to high-density neighborhoods. In the United States alone, the residential sector alone is responsible for more than 20 percent of the country's energy consumption. Thus, retrofitting buildings to improve energy efficiency, assigning higher energy standards in new developments, or releasing stricter insulation regulations will be important tools for planners to decrease energy demand in the building sector and mitigate climate change (CODEMA, 2007). A recent study by Liu and Sweeney (2012) examined the relationship between household energy demand and urban form using both an energy and land use model to compare. Using computer-generated data instead of data collected in existing neighborhoods, the results nevertheless support the findings of previous studies. Residential urban form has a significant impact on energy demand. Furthermore, type, age, and size of housing and household density are once again identified as key characteristic for reducing energy consumption for cooling or heating households. Moreover, the output of the computer models suggest that compact cities can decrease energy consumption by up to 16.2 percent per household compared to cities that are characterized by urban sprawl (Liu & Sweeney, 2012) with low-density developments.

Guided by the question if the physical form of today's cities impact private energy use at home, the study by Ko (2013) provided an extensive review of the existing literature in the fields of architecture and planning. In particular, the study examined the relationships between particular urban forms and energy use primarily for space-conditioning (heating and cooling). The urban form elements considered included (a) size and type of housing; (b) density in regards to physical compactness, dwelling units, and population; (c) community layout such as street orientation or building configuration; (d) natural elements in form of trees and other vegetation. In regards to housing type and size, the study not only confirms the results of previous research arguing that low-density settlements with predominantly detached housing use more energy for cooling and heating as multi-unit developments or attached housing. Furthermore, the author demonstrates that the increase in the average house size since 1978 has outpaced any improvement in energy efficiency over the same time span. In terms of density, research suggests that in dense populated inner city areas with mainly multifamily housing, the energy consumption per capita is lower compared to the dominant single-family housing in the suburbs (Kaza, 2010; Pitt, 2013). However, the literature review indicates that more research is needed to evaluate the potential tradeoffs of high-density developments in different climates (Ko, 2013). The aspect ratio of building height to street width can have a significant impact on energy use. On the one hand, in hot and dry conditions studies show that narrow twisting streets aligned with the usual wind direction and compactly spaced buildings with staggered heights are most energy efficient (Minne, 1988; Hough, 1995; Aggarwal, 2006). This type of aspect ratio supports natural ventilation while simultaneously diverting strong winds and enabling buildings to provide shade for each other. On the other hand, in cold regions, compact form can lead to an increase in energy use for heating due to reduced solar access (Steemers, 2003). In addition, compact urban environments can minimize heat loss from buildings and contribute to the Urban Heat Island effect (Krishan et al., 2001) as well as limit the opportunities for on-site solar energy generation (Cheng et al., 2006). Another key element effecting energy use for space conditioning addressed by Ko (2013) is the community layout. Solar neighborhood guidelines, for example, advocate an east–west street orientation, which would result in lots being oriented north–south. This would allow more south-facing buildings to maximize solar access in a neighborhood (Brown, 1985; Edminster, 2009). Nevertheless, the reviewed literature, also points out that more data are needed to quantify the impact of urban form elements on residential energy use (Ko, 2013).

When it comes to plant and surface coverage, research often points to the beneficial impacts of tress and urban parks in reducing energy consumption for heating and cooling. Tree planting efforts can improve the control of solar access, evapotranspiration, natural ventilation, and the Urban Heat Island effect (Heisler, 1986; Hildebrandt and Sarkovich, 1998; McPherson and Simpson, 2003). The climate inside parks tends to be cooler than in its urban surrounding. Studies have proven that the vegetated park is a cool patch in the built-up city. This phenomenon is often referred to as the "park cool island" effect or oasis effect (Saito et al., 1990). Observations conducted by Spronken-Smith and Oke (1999) also confirm that vegetated urban parks are often cooler than their surroundings. Various climate studies related to urban parks show the magnitude and spatial distribution of the park cool island effect. Upmanis et al. (1998) analyzed and compared parks in different cities throughout the world and validated that the cold park climate often extends beyond the park and therefore influences the temperature in surrounding urbanized areas (Upmanis et al., 1998). Yu and Hien (2006) examined the thermal benefits of city parks. In order to analyze the cooling effect of green areas, they measured temperatures at vegetated and non-vegetated locations throughout the city. Notable cooling effects of parks were reported within the vegetated urban areas, but also in the surrounding urban environment. Thus, lower temperatures in the park and in the nearby built environment prove the cooling impact of city parks. According to their research and simulations, the energy balance of an entire city can be altered by arranging green urban parks and additional evaporating surfaces throughout the whole urban settlement structure. As a result, the urban temperature will be reduced, because more heat can be dissipated (Yu & Hien, 2006). Meier (1991) also acknowledges the energy-saving potential of trees and other landscape vegetation. His study points out that vegetation can mitigate Urban Heat Islands directly by shading heat absorbing surfaces, and indirectly through evapotranspiration cooling. Furthermore, the author's research shows that vegetation consistently lowers wall surface temperatures by about 17°C and reduces air-conditioning costs by 25 to 80 percent (Meier, 1991). Moreover, urban parks support city ventilation, which is an important aspect for the mitigation of the Urban Heat Island effect, especially during the night The park breeze plays an important role in city ventilation. The theory of park breeze is based on temperature differences as a driving force for the divergent outflow of cool air at a low level (Oke, 1988).

Planning for adaptation

Even in the case of successful mitigation, impacts of climate change will remain affecting the least developed countries and poor population the hardest (IPCC, 2013). Rapid population growth in cities, especially in developing countries, will put pressure on existing infrastructure, eventually causing it to fail. This requires institutional, technical, and spatial measures to adapt to climate change impacts in urban areas where most of the world's population lives today (Biesbroek et al., 2009). Planning for adaptation is not an easy task. To address adaptation on the city level appropriately, urban planners have to accept climate change as a wicked problem (Hulme, 2009). Wicked problems lack a clear problem definition, are often interpreted differently among stakeholders, and thus a purely scientific or rational approach cannot be applied. There is no optimal solution to wicked problems, such as climate change, because of incomplete or contradictory knowledge, the number of people and opinions involved, the large economic burden, and the interconnected nature of these problems with other problems (Rittel & Weber, 1973). Global climate change is characterized by high uncertainties and the requirement and adaptation capacity for a region is often difficult to recognize. As a result, there is no single strategy at the local or (inter)national level to adapt to the impacts of climate change or reduce the GHG emissions. Instead, urban planners are confronted with high uncertainties. In terms of adaptation, the uncertainties result from the nature of climate change itself, its associated extremes, their effects, the vulnerability of systems and regions, conditions that influence vulnerability, and many attributes of adaptation, including their costs, implementability, consequences, and effectiveness (Hamin & Gurran, 2009).

Climate change adaptation also requires urban planners to more efficiently communicate with policy makers and the public in new and different ways (APA, 2008). Planners will need new communication tools to explain the possible climate change impacts and to ensure that the public and decision makers maintain the focus on long-term adaptation and mitigation responses. Engaging the public to participate in the adaptation and mitigation process is vital to its success. Many people are not aware of the precise nature, causes, and possible negative impacts of climate change (Bostrom et al., 1994; Kempton et al., 1995; Bord et al., 1998; Leiserowitz, 2010). The way the public processes information and new scientific findings regarding climate change has a significant effect on how and to what degree mitigation and adaptation strategies are supported. Climate change has to be communicated in a way that motivates the public to change their behavior and support adaptation or mitigation policies (Moser, 2006, 2010).

Spatial and urban planning presents a strategic framework in which adaptation as well as mitigation measurements can be integrated as a part of the broader perspective of sustainable development (Campbell, 2006). Due to their expertise in adapting the urban environment to the impacts of climate change, urban planners are in a position to usefully engage local stakeholders, policy makers, and decision makers to advocate for urban adaptation strategies (Corfee-Morlot, 2009).The following section examines the role of the local planner in terms of adapting the built environment to reduce the imports impacts from the potential hazards of climate change. Based on the existing literature, the questions of how planners should think and plan for adaptation as well as how planners should consider incorporating climate adaption strategies into existing and new policies are discussed. Any incorporation of climate adaption strategies should be based on in-depth impact assessments or risk analyses considering local circumstances. Generally, the practice of climate change assessments can be divided into four stages (Fuessel & Klein, 2006). The first stage focuses on the potential biophysical impacts of climate change. The next two phases consist of first and second-generation vulnerability assessments.These determine the different vulnerabilities of social groups or regions to climate change, considering their adaptive capacity.The fourth step evaluates adaptation policies and identifies specific measures to reduce the vulnerability of social groups to climate change and variability. The assessment of potential impacts of climate change on the local level supports urban planners to identify appropriate adaptation measurements (Pittock & Jones, 2000). The following paragraphs examine these critical steps, which urban planners need to consider to plan for adaption successfully.

Impact assessment

In the process of planning for adaptation, urban planners first need to consider the possible climate impacts for the particular region by primarily examining historical climate trends and climate change scenarios based on climate models. This first step is often referred to as climate impact assessment (Fuessel, 2007; IPCC, 2007; Hall, 2009).The aim is to determine how climate change could affect the natural and built-up environment of the urban region. Furthermore, a comprehensive impact assessment also identifies possible economic impacts for the city as well as social impacts on vulnerable populations (The Clean Air Partnership, 2007). Many of the possible climate change impact scenarios and the associated risks for the urban environment are based on climate models. Climate models have significantly improved over the years allowing not only global scale modeling, but also regional assessment through different downscaling techniques (Bader et al., 2008). Thus, climate models can support urban planners to reduce uncertainties and prioritize issues (Davoudi et al., 2009)

Nevertheless, there are factors that complicate these model projections and their applicability for urban planners to assess likely climate impacts. The first factor for uncertainties is the scale and resolution of the model. Since climate models still operate on a relatively large scale, they tend to forecast larger scale phenomena better than weather extremes only occurring at the local scale. For example, climate models are a valuable tool to simulate and project large scale events such as heat weaves and drought but do not capture intense rainfalls occurring only at the local level, which can lead to floods. Other important limitations are the inability of models to reproduce important atmospheric phenomena and their inaccurate representations of the complex natural interconnections (Reichler & Kim, 2008). As a result this hazard-based approach (Burton et al., 2005) of assessing climate change impacts is useful for raising awareness or identifying research priorities, but is not enough to identify priority areas for adaptation efforts (Fuessel, 2007). Instead, based on the insights and data gathered from the more general impact assessments, planners need to assess the resulting vulnerability for the urban environment in a second step.

Vulnerability assessment

Different aspects need to be addressed by planners to assess the vulnerability of built up environments to possible climate change risks and impacts (Abraham, 2009). Since the number of people living in cities is rising, vulnerabilities for urban dwellers due to climate change are increasing as well. Vulnerabilities increase due to high concentration of population in relative small area and because of socioeconomic conditions, as well as spatial characteristics in urban centers and their peripheries (UN Millennium Project, 2005). One key aspect of vulnerability assessment, especially for urban planners, is to identify vulnerable populations. Since the poorest countries of the world are those who will suffer disproportionate consequences, climate change raises issues of equity and social justice (Grasso, 2007).

Especially the urban poor faces high climate change-induced risks and vulnerabilities (Gencer, 2007). The increasing rural to urban migration have put substantial pressure on urban environments resulting in informal settlements (Weber & Puissant, 2003). Along with conditions of urban poverty, informal economy, and challenged urban management systems, these settlements and their residents have become increasingly susceptible to vulnerabilities from natural disasters. Furthermore, infrastructure deficits and uncontrolled urbanization create new hazards in informal settlements. As these settlements grow larger and denser, issues such as lack of sanitation, clean water and garbage removal, and congested living conditions add to the disaster vulnerability of slum dwellers resulting in further environmental and health problems (Haines et al., 2006). As seen in the case of hurricane Katrina, however, not only populations in poor countries are vulnerable to extreme weather events, but also the most disadvantaged in wealthy nations (Van Heerden & Bryan, 2007). Although Katrina cannot be linked directly to climate change, it represents the type of climate change impacts that we can expect in the future and thus need to be prepared for. However, in order to plan and adapt to such events appropriately equity issues need to be addressed. Planning and policy approaches need to consider local inequalities and injustice, reflected in disparities in wealth, health, education, and job opportunities. Planners are well suited to overcome these issues by identifying and addressing the root of inequities, promoting policies to reduce the problem, empowering communities, working across agencies and departments, recognizing and respecting cultural differences, and aiming for strategies for long-term permanent change.

In addition, planners need to determine to what degree specific systems such as the transportation infrastructure, the built-up environment, threatened ecosystems, or public health will be effected by the previously assessed climate impacts. Following the climate impact assessment, the first step in the vulnerability assessment process should be the evaluation of the exposure of specific systems or groups within the population to the climate change impacts (UKCIP, 2010). For example, infrastructure in coastal areas or near rivers might be exposed to flooding due to projected sea level rise or increasing precipitation. As well as in climate impact assessment, climate models play an increasing role in this part of the vulnerability assessment (Abraham, 2009). Geographical Information Systems (GIS) can produce high-resolution maps at the local scale to illustrate the exposure of urban areas to sea level rise or the Urban Heat Island effect. Such specific information leads to diminishment of uncertainties and support of urban planners in their development of appropriate adaptation strategies.

Furthermore, non-climatic factors that support or limit urban planners to manage unanticipated risks from climate change impacts have to be considered as well. Planners have a wide range of different approaches for assessing vulnerability at their disposal (Brooks, 2003; Adger, 2006; Carter et al., 2007). Scientists, governmental organizations, interest groups, and advocacy groups have developed climate adaptation planning guides. Such as the "California Climate Adaptation Planning Guide" by the California Emergency Management Agency (2012), National Oceanic and Atmospheric Administration's "Adapting to Climate Change: A Planning Guide for State Coastal Managers" (NOAA, 2010), or the Town and Country Planning Association's "climate change adaptation by design: a guide for sustainable communities" (TCPA, 2007). Although each guide differs in terms of definitions of vulnerability and how it should be determined, they all point to the three interrelated factors. These factors are exposure to impacts, sensitivity to impacts, and capacity to adapt to impacts. The sensitivity to impacts refers to the degree resources, population, infrastructure, or other important components of the urban environment respond to incremental changes in the impacts of climate change (Lowe et al., 2009). This concept allows urban planners to identify the sectors that will be influenced earliest by climate change and consequently need to be addressed first by adaptation strategies. According to the IPCC Working Group 2 (Carter et al., 2007), adaptive capacity describes the "potential or ability of a system, region, or community to adapt to the effects or impacts of climate change." On an urban scale, this means that by increasing the adaptive capacity, settlements are more likely to cope with changes and uncertainties in climate. As a result, a high adaptive capacity also increases the resiliency of cities and encourages sustainable development (Burton, 1997; Cohen et al., 1998; Smit & Wandel, 2006).

Strategy development

As mentioned above, enhancing the adaptive capacity increases the resiliency of urban areas to possible negative impacts of climate change. The idea of resiliency as a policy and planning goal has its origin in ecosystem theory (Holling, 1973) and is now commonly used in the context of climate change adaptation (Hamin & Gurran, 2009). Resiliency to climate change from a planning perspective can be understood to improve the abilities for urban systems to bounce back after suffering from a negative environmental event. Since future climate change is already built into the earth's atmosphere and short-term impacts cannot be avoided anymore, which makes adaptation a necessity (Pielke Jr. et al., 2007), it is important that planning strategies not only address the identified vulnerabilities of a city and its population, but also improve the ability to recover from negative climate change–induced events. To decrease vulnerabilities and simultaneously enhance the resiliency of cities and other urban environments, adaptation strategies should generally consider and include the following aspects:

- Improving access to resources (Ribot et al., 1996; Kelly & Adger, 1999)
- Reducing poverty (Berke, 1995; Kates, 2000)
- Lowering inequities in resources and wealth among groups (Torvanger, 1998)
- Improving education and information (Zhao, 1996)
- Improving infrastructure (Magalhaes & Glantz, 1992)
- Reducing intergenerational inequities (Munasinghe, 2000)
- Respecting accumulated local experience (Primo, 1996)
- Mitigating long-standing structural inequities (Magadza, 2000)
- Ensuring that responses are comprehensive and integrative, not just technical (Munasinghe & Swart, 2000)
- Improving institutional capacity and efficiency (Handmer et al., 1999)
- Ensuring that adaptation strategies are based on the local needs and resources (Ramakrishnan, 1999)

Still, the need to adapt to climate change is not fully acknowledged by all countries or administrations (Wamsler et al., 2013). Moreover, the countries that do recognize climate change adaptation as a significant planning challenge have only implemented very few adaptation policies (Carmin et al., 2012; Greiving & Fleischhauer, 2012; UNISDR, 2012). Overall, the important role of urban planning to the success of climate change adaption is still not recognized strongly enough among policy makers. Policy action on planning for adaptation of cities and towns is relatively new and thus many adaptation strategies are not yet translated into planning practices. Nevertheless, several adaptation strategies are already implemented in communities worldwide. According to an extensive review of current adaptation practices worldwide (Wamsler et al., 2013) that most policies that are currently proposed are quite similar and not highly focused on local circumstances. The most frequently employed measures include updating infrastructure and disaster plans to include and acknowledge projected forecasts for climate change, considering larger floodplains for areas with possible increased storm events and precipitation, establishing wildlife corridors, and adjusting building codes to support more natural cooling while contributing less to the Urban Heat Island effect (Gill et al., 2007; Hamin & Gurran, 2009). Most adaptation strategies focus on the physical structure of the built environment with a goal to reduce the vulnerability buildings and infrastructure to the impacts of climate change. The main focus by authorities in terms of urban adaptation seems to be the reduction of flood risks, landslides, extreme temperatures, urban drought and Urban Heat Island effect. In addition, preference is given to adaption policies, which also have beneficial impacts in regards to climate change mitigation and reduce GHG emissions. This is especially the case in Europe where so-called climate planning is an emerging trend which combines climate change mitigation and adaption (Davoudi et al., 2010).

A major criticism of current adaptation approaches is that most policies address physical factors separately from related non-physical factors (Wamsler et al., 2013). This means that the close interrelationships between the social, cultural, economic, political, and institutional characteristics of cities on the one side and the physical features of the urban fabric on the other are not addressed appropriately. As a result, current policy frameworks do not allow urban planning to show its full potential in terms of climate change adaption, which can lead to a further reduction in the resilience of cities instead of improving it. Depending on the local circumstances and the results from the previously performed impact and vulnerability assessments, urban planners might need to address various sectors. For example, adaptation strategies might need to consider biodiversity and habitat, infrastructure, sea level rise, public health, water resources and management, or forestry and agriculture. During the development process of adaptation strategies urban planners need to evaluate possible adaptation options according to their costs, benefits, efficiency, and implementability. They also have to be aware of potential tradeoffs and conflicts with mitigation strategies (Hamin & Gurran, 2009). For example, on the one hand, mitigation of GHG emissions at the urban scale focuses on densification of the built-up environment to reduce vehicle miles traveled and the energy use of buildings. On the other hand, adaption strategies often rely on open spaces to account for storm water, species migration, urban cooling, and other goals. Thus, planners need to develop strategies that minimize this conflict and find the right balance between minimizing the causes and impacts of climate change. Despite the fact that planners are considered well suited to address climate change risks and adaption (IPCC, 2007; Stern, 2008), it remains unclear what exactly their role is, how their responsibilities relate to the ones of city authorities, and how national adaptation policies can be translated into local planning strategies (Greiving & Fleischhauer, 2012).

Climate actions plans

To integrate climate impacts and risks into the decision-making process of urban policy and development, many cities have developed Climate Action Plans (CAP). The plans range from theoretical or motivational documents to highly detailed documents stating concrete goals with thoroughly designed methods. The first generation of Climate Action Plans focused mainly on improving municipal operations in terms of energy use and GHG emissions (Millar-Ball, 2010; Wheeler, 2009). Today, these plans are also addressing jurisdiction-wide policies such as land-use planning focusing on supporting public transit, compact development, and green building codes. In general, CAPs seem to get the most support by policy makers if the development strategies provide immediate or are highly visible results (Bassett & Shandas, 2010). By the end of the last decade, at least 141 local jurisdictions had developed CAPs (EPA, 2009). Such initiatives present a valuable instrument for urban planners to implement adaptation strategies. CAPs and their recommendations present a great opportunity to change current planning and developments patterns. CAPs can provide the framework and the political power to change current development patterns, improve the position of the planner in the decisions making process, and establish a sustainable way of living in the future, which reduces the vulnerability to climate change and increases the adaptive capacity of communities.

However, so far, planners seem to play only a small role in the development of CAPs. Instead municipalities tend to rely more on environmental, engineering, and environmental departments (Boswell et al., 2011). Furthermore only a very limited number of completed climate action plans address the need to develop and implement adaptation plans. Considering the type of policies presented in existing CAP and the lack of adaptation strategies, planners need to take a leading role in the development and implementation process. Planners have the expertise to improve CAP significantly. Many climate action plans focus on strategies that are already part of various sustainable urban planning practices, such as compact and energy saving development patterns or extensive green spaces throughout neighborhoods. Furthermore, planners have the tools necessary to improve the resiliency of urban environments and improve their adaptive capacity regarding the potential negative impacts of climate change.

Climate Action Plans present just one opportunity for urban planners to incorporate mitigation and adaptation strategies into the local planning process. Due to the uncertainties regarding future effects of climate change, adaptation measures should also be considered as part of a broader risk management strategy that takes the probabilities as well as the costs and benefits into consideration. Municipalities like Denver are addressing climate change, adaptation strategies, and mitigation measurements as part of a sustainability plan (City of Denver, 2010). Besides GHG reduction targets and strategies, the so-called Greenprint Denver initiative includes water conservation, waste diversion, economic development, land use and transportation, and natural lands management. Another approach is incorporating climate-specific urban planning into comprehensive or general plan updates. The City of San Carlos near San Francisco integrated detailed measures of a climate action plan to their general plan of broad policies (Millar-Ball, 2010). Such processes of adopting climate policies indicate that climate change has been institutionalized in local and regional government, making it easier for urban planners to implement climate adaptation strategies. These practices are often referred to as mainstreaming climate change adaptation (Klein et al., 2007; Van Aalst & Helmer, 2003), which is more thoroughly discussed in the next section.

Mainstreaming

In order for planners to actually incorporate climate adaptation strategies successfully, the issue of climate change and adaptation in particular has to be integrated or mainstreamed into existing governmental policy (Agrawala & van Aalst, 2005). The aim of mainstreaming is to make adaptation to climate change a part of other well-established programs. Otherwise, urban planners do not have the tools and the political support necessary to implement the appropriate adaptation measurements into their regular planning routine (Huq & Red, 2004). In terms of urban planning, planners need to integrate climate change risks into development policies and patterns. Thus, any decision-making process dealing with urban planning relevant issues such as urban design, water supply, and capital investments in agriculture, urban form, energy, transportation or other infrastructure should consider the impact on and resilience to climate change.

Several measurements exist to promote mainstreaming adaptation, including the integration of climate information into environmental datasets, vulnerability or hazard assessments, broad development strategies, macro policies, sector policies, or in development project design and implementation (Huq et al., 2003). Today, it is common that planned adaptations to climate change impacts and threats are developed in a way that they can be integrated into existing resource, regional, or local strategies for sustainable development. Nevertheless, there are also constraints for successfully mainstreaming climate change risks and impacts into development and urban planning. The five major constraints are relevance of climate change information for development related decisions, uncertainty of climate information, compartmentalization with governments, segmentation, and other barriers within development-cooperation agencies, and trade-offs between climate and development objectives (Agrawala & van Aalst, 2005).

Outlook: the potential to reduce emissions and increase adaptive capacities

The existing literature suggests that planning strategies have great potential in achieving significant GHG emission reduction as well as decreasing vulnerabilities and increasing adaptive capacities of urban environments. However, in regards to mitigation, the existing body of knowledge is not sufficient to come to definitive conclusions, and more research is needed either to validate the benefits of compact and transit-oriented development patterns, or to avoid unforeseen negative consequences resulting in even higher energy use and GHG emissions. As long as questions remain, reliable predictions of the impacts of land use and design strategies on travel behavior will remain elusive. The existing body of knowledge does not suggest that planning approaches implementing a development pattern based on the five *D*s are wrong-headed. Rather, it demonstrates that the success regarding the reduction of GHG emissions and energy use is not self-evident. Nevertheless, it is undeniable that the built environment is a primary contributor to climate change, current development patterns make driving a necessity in many places resulting in high energy consumption and GHG emissions. Although the impacts of land use on modal split remain unclear, planners must play a key role in promoting energy efficiency in the existing built environment and changing development patterns, transportation systems, and regulations in ways that reduce GHG emissions. At the very least, planning strategies promoting more sustainable development patterns offer various commuting choices. Furthermore, it is likely that transportation and energy costs will continue to increase considering peak oil and climate change and that more compact and walkable communities will be more resilient facing these upcoming challenges which are still characterized by large uncertainties.

Even when the potential risks and impacts are understood, the perceived long timeframes of climate change presents a significant barrier to the development and implementation of place-specific strategies. Nevertheless, more tools, resources, and ongoing efforts are becoming available to planners to provide guidance for adaptation planning. The concept of adaptation is not entirely new. Instead, it includes well-established practices from disaster risk management (e.g. early warning systems), coastal management (e.g. structural protection), resource management (e.g. water allocation), spatial planning (e.g. flood zone protection), urban planning (e.g. building codes), public health (e.g. disease surveillance), and agricultural outreach (e.g. seasonal forecasts). However, several aspects of climate change adaptation are new. Among the new challenges are unprecedented extreme climate conditions, rate of change, knowledge, methodological challenges, as well as new actors.

The diversity of adaptation challenges emphasizes the fact that it is impossible to develop a single and worldwide applicable approach for assessing, planning, and implementing adaptation measures. Any risk assessment and discussion about adaptation measurements has to allow flexible methodological approaches in order to produce knowledge that is relevant in a particular decision context. Therefore, planned adaptation to climate change means foremost the use of information about present and future climate change to examine the appropriateness of current and planned practices, policies, and infrastructure. Eventually, the role of planners is about giving recommendations regarding who should do what more, less, or differently and to determine the needed resources. Moreover, planners have to pay close attention to possible trade-offs between the considered adaption strategy and already existing mitigation policies. These two response options can sometimes be mutually reinforcing, but they can also work against each other. Thus, the appropriate adaptation strategies have to be determined on a case-by-case basis, taking the local circumstances into consideration as best as possible. Successful adaptation methods in one place cannot be fully transferred or generalized for other jurisdictions.

Policy makers, decision makers, and planners agree that mitigating GHG emissions and adapting the built environment to cope with the possible negative consequences of climate change are among the most difficult and important challenges faced by the planning profession today (Saavedre & Budd, 2009; Davoudi et al., 2010). The increasing awareness of climate change is not only more and more dictating the sustainable development debate, but is also supporting the critiques of the current predominant development patterns characterized by urban sprawl, separation of uses, and the necessity to own and operate a private automobile. The need for planners addressing alternative development strategies at different scales and different places is increasing (Liu and Sweeney, 2012). The available scientific data emphasizing the complexity, uncertainty, and irreversibility of climate change in the near future are also impacting the nature and framing of spatial planning. Planners need to be more involved in the development and implementation process of climate policies and action plans. As a result, planners will be expected to resolve or even overturn short-term and long-term expectations of development (Davoudie et al., 2010). Their work will be increasingly guided by questions such as: what will low-carbon, "climate proof," settlement look like in terms of urban form and infrastructure; what are the barriers to effective planning for such development; what are the implications for governments, from transnational to local levels and the relationship between these levels; who will bear the risk and what are the implications for equity and social development?

Furthermore, as emphasized by Bulkeley (2006), climate change is a global public good that includes complex planning issues that not only exceed the traditional framework of planning, but also the policy objectives of local authorities. One of the main issues is the limited availability of climate change impact data at the regional and local scale. Unfortunately, existing models predominantly only provide insights in terms of average changes in climate parameters at a very large geographical scales and over long time horizons. Likewise, compared to mitigation policies, which are easy to assess by measuring the change of GHG emissions over time, measuring the effectiveness of adaptation planning, which focuses on avoiding future negative impacts caused by climate change, is much more difficult. The effectiveness of adaptation strategies are still influenced by high uncertainties. More research is needed to improve the understanding of multifaceted relationships between important issues such as energy demand and consumption, land use changes, and climate change. It will take an interdisciplinary approach, in which planner will play an important role, to fully understand environmental, urban, and social problems caused by climate change.

Institutional and political barriers to successful mitigation and adaptation

Uncertainties of climate change

One of the biggest challenges for successfully developing and implementing climate change adaption and mitigation policies is the high level of uncertainties regarding the types, place, timescales, and severity of future impacts of climate change. The high levels of uncertainty derive from the different possible future behaviors and decisions by humans, from internal processes in the climate system, and from the very complex climate science itself. Especially climate change adaptation relies heavily on climate science projections compared to mitigation strategies, which focus on reducing GHG emissions in the atmosphere that are easy to measure and monitor. Adaptation strategies are based on already occurring and potential negative consequences of climate change, and decision makers need to take uncertainties into account during the strategy development process. In many cases, projected climate change impacts result in direct costs for communities, which in turn requires high investments for adaptation measures. Jurisdictions, however, struggle to commit to adaptation measures to reduce climate change impacts and improve resiliency if the potential impacts are characterized by high levels of uncertainty and low levels of probability of occurring (Boswell et al., 2011). Nevertheless, despite the fact that potential negative events are characterized by high uncertainties and low probabilities, some projected impacts are so severe and post a direct danger to communities that actions must be taken.

Framing of the climate change issue

Another considerable barrier is that climate change is usually framed as global issue with global cooperation as the main strategy against the causes and potential negative impacts. As a result, city officials and other local policy makers have often little understanding of how their decisions might impact the causes and consequences of climate change (Wilbanks & Kates, 1999). Furthermore, the global framing of the issue leads to decision makers believing that climate change is not an urgent local concern, but distant in time and space. The low level of concern and willingness to take action is further expedited by the fact that many countries, such as the United States, lack federal mandates that require municipalities to take action against the causes and impacts of climate change (Betsill, 2001). In terms of mitigation, city officials are also reluctant to take actions or commit funds to reduce GHG emissions because it cannot be guaranteed that the efforts taken locally will make any difference on the global scale. Furthermore, the location of sources emitting GHG and the area where impacts of climate change will occur are independent of each other (DeAngelo & Harvey, 1998). With more and more countries refusing to sign or extend international treaties, such as the Kyoto Protocol, local jurisdictions will be less willing to commit funds to mitigation efforts on their own.

Bureaucratic structure and institutional authority

Since the issue of climate change is very complex the causes as well as the impacts require an interdisciplinary approach with experts from different areas working together. This means that within a city government, no single department is equipped to address climate change successfully (Betsill, 2001). Instead departments responsible for transportation, waste management, health department, and others connected to the causes and impacts of climate change need to work closely together. Unfortunately, the specialized departments within a city government tend to focus only on their task and do not interact enough with other division or departments (Nijkamp & Perrels, 1994; O'Meara, 1999). This obstacle becomes very apparent in cities' efforts to control GHG emission, the main cause of anthropogenic climate change. In order to be successful in translating the political will to face the issue of GHG emissions and climate change into policies, departments responsible for waste management, transportation, public works, utilities, health, land-use planning, and air-quality management need to work together. However, many of these departments have little interaction with each other during their usual line of work and do not consider environmental protection and climate change as part of their work. As a result, the willingness to allocate financial resources as well as manpower necessary to develop or implement successful GHG emission reduction programs is limited. In addition to inadequate bureaucratic structure governments on the local level often lack the institutional authority necessary to engage effectively in climate change policies. For example legal limits, state-local relations, and lack of fiscal authority exacerbate local efforts to engage in the fight against global warming (Boswell et al., 2011). Especially constitutional barriers and local tax codes and fee authority are difficult obstacles to overcome. On the other hand, states also face legal obstacles to impose climate change policies on local jurisdictions.

Institutional capacity and funding

A different institutional barrier can have limited institutional capacity and funding. Communities often struggle to provide the means necessary to create local adaptation and mitigation programs (Betsill, 2001). This is especially the case during economic downturns and in developing countries. Compared to developed countries, communities in poorer countries lack trained personnel and face significant financial constraints, which limit access to relevant information and technologies. The lack of technology reduces the available adaptation policies and measurements to the point that nations, regions, or people cannot respond successfully to the impacts and future risks of climate change (Goklany, 1995; Burton, 1996; Iglesias et al., 1996; Scheraga & Grambsch, 1998).

Another obstacle that institutions have to address is that dealing with climate change is very time consuming and often requires additional workforce. However, many jurisdictions refuse to hire extra people and instead add the issue of climate change mitigation and adaptation to the workload of the existing and often overworked public officials. In addition, addressing climate change requires expertise and technical training to analyze complex data and to make informed decisions. This can be very cost intensive, and many city governments do not have the adequate infrastructure to obtain and access all the relevant data (Bulkeley, 2000; Kates et al., 1998). Due to the ongoing worldwide financial crises, governments on all levels tend to view environmental programs as luxury assets that take funding away from more pressing issues, such as revitalizing the local and global economy. As a result, environmental policies are often among the first programs that are cut when city revenues decrease (Press, 1998). Furthermore, climate change policies often require significant up-front investments and do not necessarily result in economic benefits.

Development and implementation process

The first barrier in the development and implementation process of mitigation and adaption policies is the impact assessment. The projected future impacts of climate change will be different by region, and communities need to determine the potential consequences for their jurisdiction individually. This, however, can be very difficult because the models used for such projection do not operate on the local scale. Instead, the various existing climate models project possible impacts mostly on the international, national, and regional scale, which need to be downscaled to the local level (Boswell et al., 2011). In addition, using historic data of environmental events to project future events is not applicable for impacts caused or aggravated by climate change. Although many impacts of climate change will still be periodic events, the severity will change and evolve over time.

Another potential barrier is the actual process of designing a policy and its implementation. Since the field of climate change research is evolving constantly, designing the correct policy is very challenging. This is especially true for adaptation strategies, which need to be flexible to be able to react to new scientific findings and impact projections. As a result, city governments need to be able to identify and gain access to data sources that allow public officials to reevaluate developed adaptation strategies regularly to ensure they still conform to the latest scientific findings. However, many decision-making frameworks within the local governments are still based on historic data and do not acknowledge the uncertain and evolving nature of climate change (Boswell et al., 2011). Furthermore, adaptation strategies need to pay close intention to possible trade-offs between the considered adaption strategy and already existing mitigation policies. These two response options can sometimes be mutually reinforcing, but they can also work against each other (IPCC, 2013). Thus, the appropriate adaptation strategies have to be determined on a case-by-case basis, taking the local circumstances into consideration as best as possible. Choosing, prioritizing, and implementing climate change strategies can be another barrier that policy and decision makers might face. Aspects such as uncertainty, multiple needs and conflicting interest, and potentially high costs make it difficult to choose and implement the correct climate change policies.

Existing forms of governance and the short-term electoral cycles

In a democratic society, the short-term electoral cycles make it very difficult for the existing forms of governance to reduce GHG emissions to the levels required for successful climate change mitigation. Due to the short-term electoral cycles, politicians are constantly concerned with their own careers, which impacts their ability to make tough policy decisions that require a large amount of political capital (Held & Hervey, 2009). Therefore, elected public officials and policy makers hesitate to introduce reforms or implement policies, which could upset voters and jeopardize their reelection (Meadowcraft, 2009). Unfortunately, effective mitigation strategies such as higher energy taxes or fuel prizes are hard to communicate to the public without losing potential voters. As a result and due to the pressure on governments to present results that can be evaluate every 4 to 5 years, policy debates tend to focus more on topics and policies that can be implemented in a short amount of time, do not require additional taxes, and can be witnessed by the public. Times of election can also hinder decisive action towards mitigation, because political leaders might be inclined to attract votes by opposing more vigorous measures that might be suggested by their opponents.

Another constraint for the existing forms of governance to achieve substantial reductions of GHG emissions is that the short-term electoral cycles are changing the political landscape constantly. New elections might result in new leadership from a party that has a different agenda regarding climate change than the previous party (Wheeler, 2009). For example, in the United States, there is a deep divide between the two main political parties, which presents a major barrier for the implementation of long-term climate change mitigation strategies. This makes it difficult to develop climate change policies when the policy makers are faced with the uncertainty of not knowing if they are still in office when the time comes to actually implement them. The two strongly different beliefs between the democratic and republican party also makes it difficult for funding initiatives or the changing of pricing structures to foster energy conservation and reduce emissions. Moreover, state and city governments are further constrained by federal preemption, which is the constitutional limitation of state power in the presence of federal power (Bushinsky, 2010). This implies that if decision makers at the federal level are opposed to addressing climate change mitigation, state and city governments who are willing to address climate mitigation will have less political and financial power to do so. This can be further amplified by the constitutional requirement that budgets proposed at the state level have to be approved by a two-thirds majority. This can give climate change mitigation critics significant veto power and limit the resources states and cities can devote to climate change (Wheeler, 2009). Furthermore, the still ongoing aftermath of the recent worldwide economic crises makes it even less likely that climate change mitigation becomes a major focus of decision makers.

Barriers and constraints such as the ones discussed, especially at the federal level, suggest that the existing democratic forms are currently too much impacted by the short-term electoral cycles to make the policy changes required to combat climate change. This emphasizes the importance of the public perception of climate change. Only if the public believes strongly in the need for mitigating climate change and is willing to commit to the required behavioral changes, elected officials will address this long-term challenge effectively.

The important role of public climate change perceptions

The existing body of knowledge already presents an important amount of possible information on planning interventions and policies needed to respond to and cope with global warming and climate change (IPCC, 2007; Newman & Beatly, 2009; NRC, 2009). Nevertheless certainly more experience and understanding is needed on institutional capacities and public attitudes as well as dealing with uncertainties for more effective responses. Designed adaptation and mitigation strategies, however, do not guarantee success in the fight against climate change (Handmer, 2003), and we do not know what needs will be, as major challenges approach highly vulnerable zones in the future. Despite the robust and convincing body for anthropogenic climate change research and science (IPCC, 2013), there is still a significant gap between the recommendations provided by the scientific community and the actual actions by the public and policy makers (Blake, 1999; Kollmuss & Agyman, 2002; Abbasi, 2006; Arvai et al., 2006). The strong findings presented by the scientific community have not transferred into long-term, comprehensive, and legally binding policy commitments especially on the national and international level. So far scientists and the media have failed to communicate the urgency of the situation successfully to policy makers and the lay public (McBean & Hengeveld, 2000; Somerville & Hassol, 2011).

As mentioned early in this chapter, existing research suggests that many people do not have a full understanding of the issues inherent in climate change. A significant part of the public is not aware of the precise nature, causes, and possible negative impacts of climate change. Despite its widespread media coverage (Bostrom et al., 1994), lay mental models of climate change suffer from several basic misconceptions (Kempton et al., 1995; Bord et al., 1998; Lorenzoni et al., 2005; Leiserowitz, 2010). Misconceptions, such as that GHG emissions are just a form of air pollution (Kempton, 1991; Brechin, 2003), result in the public support for the wrong policies. For example, many people believe and support traditional pollution controls are the solution to decrease GHG emissions. However, actions such as filters and strengthening pollution controls do not stop GHG emissions leading to climate change (Prinn et al., 2005). Climate change communication can help to advance public understanding of the issue of climate change, inform them about possible solutions, emphasize the impact of personal choices and behavior, and encourage public deliberation resulting in support for adaptation and mitigation policies and measurements (Smith, 2005; Moser, 2006; Frumkin & McMichael, 2008; Ockwell et al., 2009). However, for improved or enhanced communication broader and deeper knowledge of the public's risk perception and cultural values are needed.

The way the public processes information, how they perceive threats and other perceptional issues has a significant effect on how and to what degree mitigation and adaptation strategies are supported. Nonetheless, very little is known about public opinion and perceptions about climate change, especially at the international level (Schneider et al., 2010). One of the primary reasons for this phenomenon is the limited number of multinational surveys addressing public perceptions of climate change and its threats (Brechin, 2003; Leiserowitz, 2005, 2010). However, public risk perceptions and the understanding of climate change and public beliefs play a vital role for successfully overcoming the challenges of climate change in the next decades. In order to design, implement, and generate sufficient public support for policies and planning interventions at the national and international level, it is necessary to have a good understanding of the public's perceptions regarding climate change (Read et al., 1994; Bord et al., 1998; Moench, 2007).

The public's perceptions of climate change and the resulting behavior and degree of policy support for various mitigation options can be linked to the way the threats of climate change are communicated. Establishing accurate knowledge among the public regarding the risks, threats, and other aspects of climate change is very important and a key challenge for decision makers and communicators. Without effective communication, the public may become distrustful of the science and may not be willing to support the necessary policies to reduce GHG gasses and support adaptation investments. Yet, the field of climate change communication research, especially as a tool to change public behavior and foster public acceptance of adaptation and mitigation strategies, is still relatively young (NRC, 2010). Nevertheless, risk communication is already acknowledged as an important tool in climate change policy and research (Wardekker, 2004). We know from prior studies that when people have better understanding of climate change science, they tend to be more supportive of mitigation efforts (Read et al., 1994; Bord et al., 1998).

The purpose of the research presented in the following chapters is two-fold: first, to understand the nature of public perceptions of climate change in different countries and over time; and second, to identify perception factors which have a significant impact on the public's willingness to support climate change policies or commit to behavioral changes to reduce GHG emissions. Factors such as trust in climate change information which need to be considered in future climate change communication efforts are also addressed. The research presented and discussed contributes to the existing body of knowledge in the areas of risk perception and risk communication as well as their interrelationships, directed at climate change and strategies. Global climate change is characterized by high levels of risk, but also high levels of uncertainties. The data presented and the insights gained will permit decision makers to make better informed decisions in terms of developing, communicating, and implementing the appropriate climate change policies and strategies to successfully mitigate and adapt to the impacts of climate change.

Foundational theory

As a scholarly endeavor, we do not know very much about what risk perception factors influence the public's perceptions of climate change and if these perceptions differ from other natural threats of disasters. Therefore, this research is based on the psychometric paradigm (Fischhoff et al., 1978; Slovic et al., 1984) and seeks to advance the theories of "bounded rationality" (Simon, 1956, 1959) and "cultural cognition" of natural phenomenon (Douglas & Wildavsky, 1982). The psychometric paradigm presents a theoretical framework that implies that risk means different things to different people due to the influence of different psychological, social, institutional, and cultural factors (Slovic, 2000). This paradigm assumes through appropriate survey design, different scaling methods, and multivariate analysis to capture public risk attitudes and perceptions that are relevant to improve climate change communication programs, but fundamentally see how risk perceptions are associated with public policies on mitigation and adaptation.

The theory of bounded rationality asserts that the decision-making process of individuals is limited due to incomplete information available, cognitive limitations, and restricted amount of time to make decisions (Gigerenzer & Selten, 2001). As a result, these constraints force individuals to construct a simplified model of the world to deal with. Within the framework of bounded rationality, the key principle is the concept of "satisficing" (Slovic, 2000), which means that a person strives to attain a satisfactory solution and not necessarily an optimal level of achievement. For this research, knowledge of the workings of the public's bounded rationality regarding the complex issue of climate change allows decision makers and communicators to improve climate change communication programs and strategies. Thus, fostering behavioral change and improving the support for policies addressing the causes and potential negative impacts of climate change. Therefore, this study will test different variables that impact risk perception and risk communication.

Based on the cultural theory of risk (Douglas & Wildavsky, 1982), the theory of cultural cognition implies that individuals perceive risks according to their sense of commitment to one or another idealized form of social ordering (Thompson et al., 1990; Kahan, 2010, 2012). According to this theory, the individual's perception of the risk and threats regarding climate change is derived from and reinforced by the values they have in common with the people they are connected with. Therefore, compared to the theory of bounded rationality, the theory of cultural cognition argues that differences among the public's climate change perception are mostly caused by conflicts between opposing groups whose members' cultural outlooks dispose them to form particular perceptions (Kahan, 2010). Based on these concepts, the results of testing different aspects that impact risk perceptions and the effects of these perceptions to policy predispositions and attitudes are presented in later chapters of this book. In particular, the relationships between impacts of heuristics, trust, values, and social amplification on climate change perception and the support for various mitigation and adaptation strategies to improve risk communication efforts.

Heuristics

When laypeople are faced with the task to determine risks, they usually do not have statistical evidence on hand to base their decision on. Instead, they rely on assumptions based on what they remember hearing or observing about the risk they are confronted with (Slovic, 1987). As pointed out by Short (1984), social influences impacting the response to hazards are mainly transmitted by friends, family, fellow workers, and respected public officials media. Moreover, since the 1980s, researchers, especially in the field of psychology, were able to identify various general rules that guide people in forming their perception. Known as heuristics, these judgment rules are applied by laypeople to reduce difficult mental tasks to simpler ones (Kahneman et al., 1982; Makofske & Edelstein, 1988).

The heuristic known as the *availability heuristic* is very important for the formation of risk perceptions (Tversky & Kahneman, 1973). By applying this heuristic, people judge an event as probable or frequent if instances of it are easy to imagine or remember. Since events that happen more often are usually easier to imagine and recall than unusual, rare, events, the availability heuristic is often an appropriate cue. However, other factors such as a recent disaster can affect "availability" and thus distort risk judgments. Several studies identified errors caused by using this heuristic (Lichtenstein et al., 1978). Research demonstrates that the people's judgments are moderately accurate in a global sense, but there is also evidence that shows serious misjudgments reflecting the availability bias. For example, rare causes of death are often overestimated by lay people, and common causes of death are underestimated. Another example, discussed in the study by Lichtenstein et al. (1978) is that homicides were perceived more frequently than diabetes or stomach cancer. These biasing effects of memorability and imaginability present a barrier to open, objective discussion of risk (Slovic, 2000).

Other significant heuristics are the *confirmation heuristic* and the *overconfidence heuristic*. Once a person forms an opinion, the confirmation heuristic can result in a situation where new evidence is misinterpreted or altered in order to support the initial conclusion. For people who apply the confirmation heuristic, new evidence regarding the issue of global climate change only will appear reliable if it is consistent with one's initial beliefs (Slovic et al., 1984). For example, ambiguous data may be interpreted as a confirmation. Furthermore, contrary evidence may be filtered out because they are perceived as unreliable, erroneous, or unrepresentative. The overconfidence heuristic, on the other hand, suggests that people often have too much confidence in their own judgments (Slovic et al., 1981). Research suggests that overconfidence can prevent the public to realize how little they know about climate change and how much additional information they need regarding the risks, threats, and possible adaptation and mitigation strategies (Slovic, 2000).

Thus, if the heuristics are invalid for the risk faced they can lead to large and persistent biases, thus impacting public risk assessment, judgment and policy preferences. This study will advance this theory by examining those perceptual factors that influence the public's risk perception of climate change and affect the level of awareness and concern, personal behavior, and climate policy support. Another theory scrutinized is that lay people dealing with uncertainties tend to over- or underrate the risks and threats (Slovic, 2000). In the case of nuclear power risk perception, research shows that the public tends to overrate the risks of radiation leading to large social amplification effects and behavior. The data presented in this book tested if the lay public also over- or underrates the risks and possible impacts of climate change. The overrating of climate change risks may result in a pattern of policy preferences different than responses for underrated hazard risks. The study will test these relationships.

Trust

The role of trust is another important aspect that influences risk perception (Van de Vusse, 1993; Slovic, 1997). Trust is a multi-faceted concept that includes cognitive, affective, and behavioral dimensions (Bradbury et al., 1999). Moreover, trust is also a dynamic process taking place at the individual, the institutional, and the ideological level (Tait, 2011). Besides building trust through interpersonal relations, people can also hold trust in organizations and institutions (Hardin, 2006), or in ideological values and norms (Luhmann, 1979; Blackburn, 1998). These layers in which trust operates are not mutually exclusive.

To date, various risk communication programs in Europe and the United States only show limited effectiveness (Cvetkovich & Loefstedt, 1999). Research shows that the failure of risk communication are significantly influenced by the public's trust in the communicator and in the ability of certain individuals, industries, or institutions responsible for risk management (Renn & Levine, 1991; Kasperson et al., 1992; Nye Jr. et al., 1997). In most circumstances, new information is first judged based on the credibility of its source. If there is no trust in the source, any message is likely to be disregarded, no matter how well intentioned and well delivered. Impacting the level of trust towards risk communicators and risk managers are factors such as perceived competence, objectivity, fairness, consistency, and goodwill. Especially in areas characterized by high uncertainties, as in climate change, trust plays a vital role in the success of risk communication programs and implementation of policies. Moreover, trust is not only a necessary precondition for successful climate change communication, but it can also be improved by well–developed communication strategies (Misztal, 1996). Trust in organizations whose risk management policies impact communities and the environment is vital in order to reduce complexity and generate social cooperation.

Therefore, the theory tested in the context of climate change is that distrust of certain individuals, industries, scientists, or institutions responsible for climate change risk management is strongly linked to the level of risk perceived (Mushkatel & Pijawka, 1992).

Social values

Social values are another aspect that impacts risk perception as well as risk acceptance (Slovic, 1987). The important role of social values became apparent when studies of risk perception showed that exaggerated public concern was not just a result of the public's ignorance or irrationality (Slovic, 2000). Instead, the public's reaction to risk could be linked to sensitivity to technical, social, and psychological qualities of hazards that were not well or at all communicated in technical risk assessments. For example, qualities such as uncertainty in risk assessments, perceived inequity in the distribution of risks and benefits, and aversion to being exposed to risk that were involuntary, not under one's control, or dreaded. Cultural theorists argue that our worldviews and our values play an important role in public risk perception and behavior (Douglas et al., 1998). Thus, worldviews and values have a strong impact on how the risk and threats of climate change are perceived and to what degree different strategies are supported (Hulme, 2009). For instance, members of the Republican Party in the United States tend to hold more conservative values compared to their counterparts in the US Democratic Party. As a result, republicans are often more skeptical towards the concept of human induced climate change and view policy measures as regulatory burdens and thus are less likely to support any climate change policies.

Social amplification

Another aspect complicating how people perceive, evaluate, and act on climate change risk is "the social amplification of risk" concept (Hulme, 2009). Social amplification of risk (Kasperson et al., 1988) implies that risks are communicated through different signals such as images, signs, and symbols. By interacting with psychological, institutional, or cultural processes in society, these signals can amplify or attenuate the perception of risks and their manageability. The public is embedded in this complex web of interactions where risks are symbolized, translated, and interpreted in numerous ways and by multiple actors. The research presented advances the theory of social amplification by determining where climate change is positioned on the hierarchy of environmental hazards in terms of its potential for social amplification.

Theoretical framework

In order to explore the linkages between public perceptions of climate change and personal behavior, a theoretical framework was developed based on the existing body of literature. As shown in Figure 2.1, climate change requires policy makers to continuously develop and implement adaptation and mitigation policies. In order to implement these strategies successfully the risks of climate change and the responding policies need to be communicated to a public in a way that ensures their support. As noted before, it is crucial for the success of any climate change policy that the public supports it and is willing to commit to behavioral and policy changes. Nevertheless, communication efforts can only be successful if they incorporate the factors in public risk perceptions. Therefore, the created theoretical framework supports the argument that different mediating factors impact the public's perceptions towards climate change communication and risk. These factors are categorized into the three different groups – Global Climate Change Events, Psychometric Factors, and Uncertainty of Global Climate Change. They are displayed in the center of Figure 2.1.

As illustrated in the upper right corner of the theoretical framework, in addition to the mediating factors individuals' own risk assessment is also a factor that impacts the personal perception of climate change risks and the successfulness of communication programs. Based on the perceived risk, the public assesses the personal risk resulting from climate change and decides whether or not they are willing to support important climate policies and change their behavior. Nevertheless, even the best-designed communication programs, based on the strongest social and decision science produce only best guesses about how

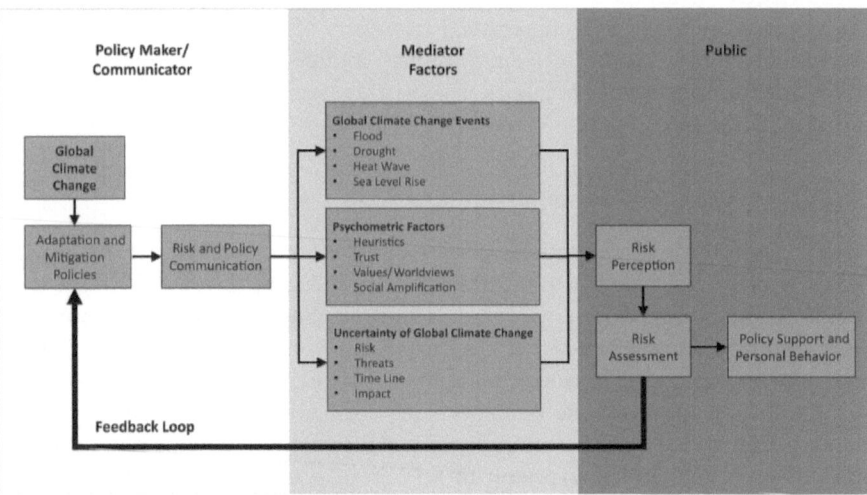

Figure 2.1 Theoretical framework

Source: Author's illustration.

to formulate messages (Pidgeon & Fischhoff, 2011). Thus, empirical testing is always needed in order to determine how effective the current communication strategy is (Moser, 2010). This is represented by the feedback loop shown at the bottom of Figure 2.1. Yet, despite the critical importance of climate change communication, such evaluations are remarkably rare. Instead, most communications rely on intuitive notions of what to say and how to say it. A scientific approach to communication science, however, requires the systematic feedback provided by empirical evaluation.

In order to test the mediating factors and support the validity of the theoretical framework, the data presented are from an international public survey of nine countries. The nine countries include the United States, Canada, Mexico, Brazil, Spain, Germany, the United Kingdom, the Netherlands, and Japan. The collected survey data allow analyzing the public's risk perceptions of climate change, their willingness to support climate policies, and their readiness to commit to behavioral changes. This provides vital information for the feedback loop, which supports policy makers and communicators to evaluate whether or not implemented policies and communication programs are successful or need to be improved. The survey instruments as well as the other research methods applied are discussed in detail in Chapter 3, which focuses on the methodology of the presented research.

Bibliography

Abbasi, D.R. (2006). *Americans and Climate Change: Closing the Gap between Science and Action.* Yale University, CT: School of Forestry & Environmental Studies.

Abraham, J. (2009). *Southwest Climate Change Network: Assessing Climate Impacts.* Tucson: The University of Arizona. http://www.southwestclimatechange.org/solutions/adaptation/assessing-impacts (accessed 10/4/2014)

Adger, W.N. (2006). Vulnerability. *Global Environmental Change, 16*(3), 268–271.

Aggarwal, R. (2006). *Energy Design Strategies for City-centers: An Evaluation.* Paper presented at the 23rd Conference on Passive and Low Energy Architecture, Geneva, Switzerland, September 6–8.

Agrawala, S., & van Aalst, M. (2005). Bridging the gap between climate change and development. In S. Agrawala (Ed.), *Bridge Over Troubled Waters: Linking Climate Change and Development* (pp. 133–146). Paris, France: Organisation for Economic Co-operation and Development.

Alexander, D., & Tomalty, R. (2002). Smart growth and sustainable development: Challenges, solutions and policy directions. *Local Environment, 7*(4), 397–409.

American Planning Association (APA). (2008). *Policy Guide on Planning & Climate Change.* Washington, DC: Author.

American Society of Planning Officials (ASPO). (1951). *The Journey to Work: Relation Between Employment and Residence.* Planning Advisory Service Report no. 26. Chicago, IL: Author.

Anderson, W.P., Kanaroglou, P.S., & Miller, E.J. (1996). Urban form, energy and the environment. A review of issues, evidence and policy. *Urban Studies, 33*(1), 7–36.

Arvai, J., Bridge, G., Dolsak, N., Franzese, R., Koontz, T., & Luginbuhl, A. (2006). Adaptive management of the global climate problem: Bridging the gap between climate research and climate policy. *Climatic Change, 78*(1), 217–225.

Bader, D.C., Covey, C., Gutowski, W.J., Held, L.M., Kunkel, K.E., Miller, R.L., . . . Zhang, M.H. (2008). *Climate Models: An Assessment of Strengths and Limitations.* Washington, DC: Department of Energy.

Bassett, E., & Shandas, V. (2010). Innovation and climate action planning. *Journal of the American Planning Association, 76*(4), 435–450.

Berke, R.P. (1995). Natural-hazard reduction and sustainable development: A global assessment. *Journal of Planning Literature, 9*(4), 370–382.

Berman, M.A. (1996). The transportation effects of neo-traditional development. *Journal of Planning Literature, 10*(4), 347–363.

Betsill. M. (2001). Mitigation climate change in US cities: Opportunities and obstacles. *Local Environment, 6*(4), 393–406.

Biesbroek, G.R., Swart, R.J., & van der Knaap, W.G.N. (2009). The mitigation-adaptation dichotomy and the role of spatial planning. *Habitat International, 33,* 230–237.

Blackburn, S. (1998). Trust, cooperation, and human psychology. In V. Braithwaite & M. Levi (Eds.), *Trust and Governance* (pp. 28–45). New York, NY: Russell Sage Foundation.

Blake, J. (1999). Overcoming the "Value-Action Gap" in environmental policy: Tensions between national policy and local experience. *Local Environment, 4*(3), 257–278.

Boarnet, M. G, & Crane, R. (2001). *Travel by Design: The Influence of Urban Form.* New York, NY: Oxford University Press.

Boer, R., Zheng, Y., Overton, A., Ridgeway, G.K., & Cohen, D.A. (2007). Neighborhood design and walking trips in ten U.S. metropolitan areas. *American Journal of Preventive Medicine, 32*(4), 298–304.

Bord, R.J., Fisher, A., & O'Connor, R.E. (1998). Public perceptions of global warming: United States and international perspectives. *Climate Research, 11,* 75–84.

Bostrom, A., Granger Morgan, M., Fischhoff, B., & Read, D. (1994). What do people know about global climate change? *Risk Analysis, 14*(6), 959–970.

Boswell, M.R., Greve, A.I., & Tammy, L.S. (2011). *Local Climate Action Planning.* Washington, DC: Island Press.

Bradbury, J.A., Branch, K.M., & Focht, W. (1999). Trust and public participation in risk policy issues. In G. Cvetkovich & R.E. Loefstedt (Eds.), *Social Trust and the Management of Risk* (pp. 117–127). London, UK: Earthscan Publications Ltd.

Brechin, S.R. (2003). Comparative public opinion and knowledge on global climate change and the Kyoto Protocol: The U.S. versus the world? *International Journal of Sociology and Social Policy, 23*(10), 106–34.

Brooks, N. (2003). *Vulnerability, Risk and Adaptation: A Conceptual Framework.* University of East Anglia, UK: Tyndall Center for Climate Change Research.

Brown, G.Z. (1985). *Sun, Wind, and Light: Architectural Design Strategies.* New York, NY: John Wiley.

Bulkeley, H. (2000). Down to earth: Local government and greenhouse policy in Australia. *Australian Geographer, 31*(3), 289–308.

Bulkeley, H. (2006). A changing climate for spatial planning. *Planning Theory & Practice, 33*(2), 203–214.

Bulkeley, H., & Betsill, M. (Eds.). (2003). *Cities and Climate Change: Urban Sustainability and Global Environmental Governance.* London, UK: Routledge

Burton, I. (1996). The growth of adaptation capacity: Practice and policy. In J. Smith, N. Bhatti, G. Menzhulin, R. Benioff, M.I. Budyko, M. Campos, . . . F. Rijsberman (Eds.), *Adapting to Climate change: An International Perspective* (pp. 55–67). New York, NY: Springer Verlag.

Burton. I. (1997). Vulnerability and adaptive response in the context of climate and climate change. *Climatic Change, 36,* 185–196.

Burton, I., Malone, E., Huq S., Lim, B., & Spanger-Siegfried, E. (2005). *Adaptation Policy Frameworks for Climate Change: Developing Strategies, Policies and Measures.* Cambridge, UK: Cambridge University Press.

Bushinsky. J. (2010). U.S. state climate action. In S.H. Schneider, A. Rosencranz, M.D. Mastrandrea, K. Kuntz-Duriseti (Eds.), *Climate Change Science and Policy* (pp. 371–376). Washington: Island Press.

California Emergency Management Agency. (2012). *California Climate Adaptation Planning Guide.* http://resources.ca.gov/climate_adaptation/local_government/adaptation_planning_guide.html (last accessed 3/20/14)

Calthorpe, P. (1993). *The Next American Metropolis: Ecology, Community, and the American Dream.* New York, NY: Princeton Architectural Press.

Calthorpe, P. (2010). *Urbanism in the Age of Climate Change.* Washington, DC: Island Press.

Campbell, H. (2006). Is the issue of climate change too big for spatial planning? *Panning Theory and Practice, 7*(2), 201–203.

Carmin, J., Nadkarni, N., & Rhie, C. (2012). *Progress and Challenges in Urban Climate Adaptation Planning: Results of a Global Survey.* Cambridge, MA: MIT.

Carter, T.R., Mearns, L.O., Conde, C., O'Neill, B.C., Roundsevell, M.D.A., Zurek, M.B., . . . Bhadwal, S. (2007). New assessment methods and the characterization of future conditions. In *Climate Change 2007: Impacts, Adaptation and Vulnerability. Contribution of Working Group II to the Fourth Assessment Report of the Intergovernmental Panel on Climate Change.* Cambridge, UK: Cambridge University Press.

Cervero, R., & Gorham, R. (1995). Commuting in transit versus automobile neighborhoods. *Journal of the American Planning Association, 61*(2), 210–225.

Cheng, V., Steemers, K., Montavon, M., & Compagnon, R. (2006). *Urban Form, Density and Solar Potential.* Paper presented at the 23rd Conference on Passive and Low Energy Architecture, Geneva, Switzerland, September 6–8.

City of Denver. (2010). *Greenprint Denver: Building a sustainable city together, today.* http://www.greenprintdenver.org/about/climate-action-plan-reports (accessed 3/24/2014)

The Clean Air Partnership. (2007). *Cities Preparing for Climate Change: A Study of Six Urban Regions.* Toronto, Canada: Clean Air Partnership.

CODEMA. (2007). *Progress on the Development of an Action Plan on Energy for Dublin.* Dublin, Ireland: Dublin City Council.

Cohen, S., Demeritt, D., Robinson, J., & Rothman, D. (1998). Climate change and sustainable development: Towards dialogue. *Global Environmental Change, 8*(4), 341–371.

Corfee-Morlot, J. (2009), *California in the Greenhouse: Regional Climate Change Policies and the Global Environment*, PhD dissertation, Geography Department. London, UK: University College London.

Crane, R. (1996). Cars and drivers in the new suburbs: Linking access to travel in neotraditional planning. *Journal of the American Planning Association, 62*(1), 51.

Crane, R. (2000). The influence of urban form on travel: An interpretive review. *Journal of Planning Literature, 15*(1), 3–23.

Crane, R., & Crepeau, R. (1998). Does neighborhood design influence travel? A behavioral analysis of travel diary and GIS data. *Transportation Research Part D: Transport and Environment, 3*(4), 225–238,

Cvetkovich, G., & Loefstedt, R. (Eds.). (1999). *Social Trust and the Management of Risk.* London, UK: Earthscan.

Davoudi, D., Crawford, J., & Mehmood, A. (2009). Climate change and spatial planning responses. In S. Davoudi, J. Crawford, & A. Mehmood (Eds.), *Planning for Climate Change: Strategies for Mitigation and Adaptation for Spatial Planners* (pp. 7–18). London, UK: Earthscan.

Davoudi, D., Crawford, J., & Mehmood, A. (Eds.). (2010). *Planning for Climate Change: Strategies for Mitigation and Adaptation for Spatial Planners.* London, UK: Earthscan.

DeAngelo, B., & Harvey, L.D. (1998). The jurisdictional framework for municipal action to reduce greenhouse gas emissions: Case studies from Canada, USA and Germany. *Local Environment, 3*(2), 111–136.

De la Barra, T. (1989). Integrated land use and transport modeling: Decision chains and hierarchies. *Cambridge Urban and Architectural Series*, no. 12. Cambridge, UK: Cambridge University Press.

DEFRA (Department for Environment, Food & Rural Affairs). (2005). Securing the Future: Delivering UK Sustainable Development Strategy. http://www.sustainabledevelopment. gov.uk/publications/uk-strategy-2005.htm (accessed 8/8/2014)

Douglas, M., Gasper, D., Ney, S., & Thompson, M. (1998). Human needs and wants. In S. Rayner and E.L. Malone (Eds.), *Human Choice and Climate Change* (pp. 195–265). Columbus, OH: Battelle Press.

Douglas, M., & Wildavsky, A. (1982). *Risk and Culture: An Essay on the Selection of Technological and Environmental Dangers.* Berkeley: University of California Press.

Edminster, A.V. (2009). *Energy Free: Homes for a Small Planet.* San Rafael, CA: Green Building Press.

Ellis, C. (2002). The new urbanism: Critiques and rebuttals. *Journal of Urban Design, 7*(3), 261–291.

Environmental Protection Agency (EPA). (2009). Climate Change Action Plans. http://www. epa.gov/statelocalclimate/local/local-examples/action-plans.html (accessed 10/2/2014)

Ewing, R., Bartholomew, K., Winkelman, S., Walters, J., & Anderson, G. (2008a). Urban development and climate change. *Journal of Urbanism: International Research on Placemaking and Urban Sustainability, 1*(3), 201–216.

Ewing, R., Bartholomew K., Winkelman, S., Walters, J., & Chen, D. (2007). *Growing Cooler: Evidence on urban Development and Climate Change.* Chicago, IL: Urban Land Institute.

Ewing R., Nelson, A.C., & Bartholomew, K. (2009). *Response to Special Report 298 Driving and the Built Environment. Unpublished Comment.* Salt Lake City: University of Utah.

Ewing, R., & Rong, F. (2008). The impact of urban form on U.S. residential energy use. *Housing Policy Debate, 19*(1), 1–30.

Fischhoff, B., Slovic, P., Lichtenstein, S., Read, S., & Combs, B. (1978). How safe is safe enough? A psychometric study of attitudes towards technological risks and benefits. *Policy Sciences, 9*, 127–152.

Frumkin, H., & McMichael, A.J. (2008). Climate change and public health: Thinking, communicating, acting. *American Journal of Preventive Medicine, 35*(5), 403–410.

Fuessel, H.M. (2007). Adaptation planning for climate change: Concepts, assessment approaches, and key lessons. *Sustainability Science, 2*, 265–275.

Fuessel, H.M., & Klein, R.J.Y. (2006). Climate change vulnerability assessments: An evolution of conceptual thinking. *Climate Change, 75*(3), 301–329.

Gencer, E.A. (2007). *Vulnerability in Hazard-Prone Megacities: An Overview of Global Trends and the Case of the Istanbul Metropolitan Area.* Summer Academy for Social Vulnerability.

Gigerenzer, G., & Selten, R. (Eds.). (2001). *Bounded Rationality: The Adaptive Toolbox.* Cambridge, MA: MIT Press.

Gill, S.E., Handely, J. F, Ennos, A.R., & Pauleit, S. (2007). Adapting cities for climate change: The role of green infrastructure. *Built Environment, 33*(1), 115–133.

Goklany, I.M. (1995). Strategies to enhance adaptability: Technological change, sustainable growth and free trade. *Climatic Change, 30*, 427–449.

Grasso, M. (2007). A normative ethical framework in climate change. *Climate Change, 81*, 223–246.

Greiving, S., & Fleischhauer, M. (2008). Raumplanung: in Zeiten des Klimawandels wichtiger denn je! Groessere Planungsflexibilitaet durch informelle Ansaetze einer Klimarisiko-Governance. *Raumplanung, 137*, 61–66.

Greiving, S., & Fleischhauer, M. (2012). National climate change adaptation strategies of European states from a spatial planning and development perspective. *European Planning Studies, 20*, 27–48.

Haines, A., Kovats, R.S., Campbell-Lendrum, D., & Corvalan, C. (2006). Climate change and human health: Impacts vulnerability and public health. *Public Health, 120*, 585–596.

Hall, N.D. (2009). Interstate environmental impact assessments. *Environmental Law Reporter, 39*(7), 9–19.

Hallegatte, S., Henriet, F., & Corfee-Morlot, J. (2008). *The Economics of Climate Change Impacts and Policy Benefits at City Scale: A Conceptual Framework*. OECD Environment Working Paper 4, ENV/WKP(2008)3, Paris, France: Organisation for Economic Co-operation and Development.

Hamin, E.M., & Gurran, N. (2009). Urban Form and climate change: Balancing adaptation and mitigation in the U.S. and Australia. *Habitat International, 33*, 238–245.

Handmer, J. (2003). Adaptive capacity: What does it mean in the context of natural hazards? In J.B. Smith, R.J.T. Klein, & S. Huq (Eds.), *Climate Change, Adaptive Capacity and Development* (pp. 51–70). London, UK: Imperial College Press.

Handmer, J., Dovers, S., & Downing, T.E. (1999). Societal vulnerability to climate change and variability. *Mitigation and Adaptation Strategies for Global Change, 3*(3–4), 267–281.

Handy, S. (2005). Smart growth and the transportation-land use connection: What does the research tell us? *International Regional Science Review, 28*(2), 146–167.

Handy, S.L., & Clifton, K.J. (2001). Local shopping as a strategy for reducing automobile travel. *Transportation, 28*(4), 317–346.

Hardin, R. (2006). *Trust*. Cambridge, UK: Policy Press.

The Heinz Center. (2007). *A Survey of Climate Change Adaptation Planning*. Washington, DC: The H. John Heinz III Center for Science, Economics and the Environment.

Heisler, G.M. (1986). Energy savings with trees. *Journal of Arboriculture, 12*(5), 113–125.

Held, D., & Hervey, A.F. (2009). *Democracy, Climate Change and Global Governance: Democratic Agency and the Policy Menu Ahead*. London, UK: Policy Network.

Hildebrandt, E.W., & Sarkovich, M. (1998). Assessing the cost-effectiveness of SMUD's Shade Tree Program. *Atmospheric Environment, 32*(1), 85–94.

Holden, E., & Norland, I.T. (2005). Three challenges for the compact city as a sustainable urban form: Household consumption of energy and transport in eight residential areas in the Greater Oslo Region. *Urban Studies, 42*(12), 2145–2166.

Holling, C.S. (1973). Resilience and stability of ecological systems. *Annual Review of Ecology and Systematics, 4*, 1–23.

Hough, M. (1995). Cities and Natural Process. London, UK: Routledge.

Hulme, M. (2009). *Why We Disagree About Climate Change: Understanding Controversy Inaction and Opportunity*. New York, NY: Cambridge University Press.

Huq. S, Rahman, A A., Konate, M., Sokona, Y., & Reid, H. (2003). *Mainstreaming Adaptation to Climate Change in Least Developed Countries (LDCS)*. London, UK: International Institute for Environment and Development.

Huq, S., & Red, H. (2004). Mainstreaming adaptation in development. *IDS Bulletin. 35*(3), 15–21.

Iglesias, A.L., Erda, L., & Rosenzweig, C. (1996). Climate change in Asia: A review of the vulnerability and adaptation of crop production. *Water, Air, and Soil Pollution, 92*, 13–27.

Intergovernmental Panel on Climate Change (IPCC). (2007). *Climate Change 2007: The Physical Science Basis. Contribution of Working Group I to the Fourth Assessment Report of the Intergovernmental Panel on Climate Change.* Cambridge, UK/New York, NY: Cambridge University Press

IPCC. (2013). *Climate Change 2013: The Physical Science Basis. Contribution of Working Group I to the Fifth Assessment Report of the Intergovernmental Panel on Climate Change.* Cambridge, UK/New York, NY: Cambridge University Press.

IPCC. (2014). *Climate Change 2014: Mitigation of Climate Change. Contribution of the Working Group 2 to the Fifth Assessment Report of the Intergovernmental Panel on Climate Change.* Cambridge, UK/New York, NY: Cambridge University Press.

Kahan, D.M. (2010). Fixing the communications failure. *Nature, 463*, 296–297.

Kahan, D.M. (2012). Cultural cognition as a conception of the cultural theory of risk. In S. Roeser, R. Hillerbrand, & M. Peterson (Eds.). *Handbook of Risk Theory: Epistemology, Decision Theory, Ethic and Social Implications of Risk* (pp. 726–759). London, UK: Springer.

Kahneman, D., Slovic, P., & Tversky, A. (Eds.). (1982). *Judgment Under Uncertainty: Heuristics and Biases.* New York, NY: Cambridge University Press.

Kasperson, R., Golding, D., & Tuler, S. (1992). Social distrust as a factor in sitting hazardous facilities and communicating risks. *Journal of Social Issues, 48*(4), 161–187.

Kasperson, R.E., Renn, O., Slovic, P., Brown, H.S., Emel, J., Goble, R., . . . Ratick, S. (1988). The social amplification of risk: A conceptual framework. *Risk Analysis, 8*(2), 177–187.

Kates, R.W. (2000). Cautionary tales: adaptation and the global poor. *Climatic Change, 45*(1), 5–17.

Kates, R.W., Mayfield, M.W., Torrie, R.D., & Witcher, B. (1998) Methods for estimating greenhouse gases from local places. *Local Environment, 3*(3), pp. 279–297.

Kaza, N. (2010). Understanding the spectrum of residential energy consumption: A quantile regression approach. *Energy Policy, 38*(11), 6574–6585.

Kelly, E. (1994). The transportation landuse link. *Journal of Planning Literature, 9*(2), 128–145.

Kelly, P., & Adger, W.N. (1999). *Assessing Vulnerability to Climate Change and Facilitation Adaptation. Working Paper GEC 99–07.* Norwich, UK: University of East Anglia.

Kempton, W. (1991). Lay perspectives on global climate change. *Global Environmental Change, 1*(3), 183–208.

Kempton, W., Boster, J, & Hartley, J. (1995). *Environmental Values in American Culture.* Cambridge, MA: The MIT Press.

Khattak, A.J., & Rodriguez, D. (2005). Travel behavior in neotraditional neighborhood developments: A case study in USA. Transportation Research Part A. *Policy and Practice, 39*(6), 481–500.

Klein, R.J.T., Erikson, S.E.H., Naess, L.O., Hammill, A., Tanner, T.M., Robledo, C., & O'Brian, K.L. (2007). Portfolio screening to support the mainstreaming of adaptation to climate change into development assistance. *Climatic Change, 84*, 23–44.

Ko, Y. (2013). Urban form and residential energy use: A Review of design principles and research findings. *Journal of Planning Literature, 28*(4), 327–351.

Kollmuss, A., & Agyeman, J. (2002). Mind the gap: Why do people act environmentally and what are the barriers to pro-environmental behavior? *Environmental Education Research, 8*(2), 239–260.

Krishan, A., Baker, N., Yannas, S., & Szokolay, S. (2001). *Climate Responsive Architecture: A Design Handbook for Energy Efficient Buildings*. New Delhi, India: Tata McGraw-Hill.

Leiserowitz, A. (2005). American risk perceptions: Is climate change dangerous? *Risk Analysis, 25*(6), 1433–1442.

Leiserowitz, A. (2010). Risk perception and behavior. In S. H. Schneider, A. Rosencranz, M.D., Mastrandrea, & K. Kuntz-Duriseti (Eds.), *Climate Change Science and Policy* (pp. 175–184). Washington, DC: Island Press.

Lichtenstein, S., Slovic, P., Fischhoff, B., Layman, M., & Combs, B. (1978). Judged frequency of lethal events. *Journal of Experimental Psychology: Human Learning and Memory, 4*, 551–578.

Liu, X., & Sweeney, J. (2012). Modelling the impacts of urban form on household energy demand and related CO2 emissions in the Greater Dublin Region. *Energy Policy, 46*, 359–369.

Lorenzoni, I., Pidgeon, N., & O'Connor, R. (2005). Dangerous climate change: The role for risk research. *Risk Analysis, 25*(6), 1387–1398.

Lowe, A., Foster, J., & Winkelman, S. (2009). *Asking the Climate Question: Adapting to Climate Change Impacts in Urban Regions*. Washington, DC: Center for Clean Air Policy.

Luhmann, N. (1979). *Trust and Power*. Chichester, UK: Wiley.

Lund, H. (2006). Reasons for living in a transit-oriented development, and associated transit use. *Journal of the American Planning Association, 72*(3), 357–366.

Magadza, C.H.D. (2000). Climate change impacts and human settlements in Africa: Prospects for adaptation. *Environmental Monitoring and Assessment, 61*(1), 193–205.

Magalhaes, A.R., & Glantz, M.H. (1992). *Socio-Economic Impacts of Climate Variations and Policy Responses in Brazil*. Brasilia, Brazil: United Nations Environment Program.

Makofske, J.W., & Edelstein, M.R. (1988). *Radon and the Environment*. Park Ridge, IL: Elsevier Science.

McBean, G.A., & Henegveld, H.G. (2000). Communicating the science of climate change: A mutual challenge for scientists and educators. *Canadian Journal of Environmental Education, 5*(1), 9–23.

McEvoy, D., Lindley S., & Handley, J. (2006). Adaptation and mitigation in urban areas: Synergies and conflicts. *Proceedings of the Institution of Civil Engineers – Municipal Engineer, 159*(4), 185–191.

McPherson, G.E., & Simpson, J.R. (2003). Potential energy savings in buildings by an urban tree planting programme in California. *Urban Forestry & Urban Greening, 2*(2), 73–86.

Meadowcroft, J. (2009). *Climate Change Governance. Background Paper to the 2010 World Development Report*. Washington, DC: The World Bank Development Economics World Development Report Team.

Meier, A.K. (1991). Measured cooling savings from vegetative landscaping. *Energy Efficiency and the Environment, 4*(3), 133–143.

Millar-Ball, A. (2010). Where the action is. *Planning, 76*(7), 17–21.

Minne, A. (1988). *Energy Design Principles in Buildings*. Louvain-la-Neuve, Belgium: International Energy Agency: Solar Heating and Cooling Program.

Misztal, B.A. (1996). *Trust in Modern Societies*. Cambridge, UK: Polity Press.

Moench, M. (2007). Adapting to climate change and the risks associated with other natural hazards: Methods for moving from concepts to action. In M. Moench & A. Dixit (Eds.), *Working with the Winds of Change*. Kathmandu, Nepal: ISET-Nepal.

Moser, S.C. (2006). Talk of the city: Engaging urbanities on climate change. *Environmental Research Letters, 1*, 1–10.

Moser, S.C. (2010). Communicating climate change: history, challenges, process and future directions. *Wiley Interdisciplinary Reviews: Climate Change, 1*(1), 31–53.

Munasinghe, M. (2000). *Development, equity and sustainability (DES) in the context of climate change*. Proceedings of the IPCC Expert Meeting held in Colombo, Sri Lanka.

Munasinghe, M., & Swart, R. (2000). *Climate change and its linkages with development, equity, and sustainability*. Proceedings of the IPCC Expert Meeting held in Colombo, Sri Lanka.

Mushkatel, A.H., & Pijawka, K.D. (1992). *Institutional Trust, Information and Risk Perceptions: Report of Findings of the Las Vegas Metropolitan Area Survey*. Carson City: Nevada Nuclear Waste Project Office.

National Oceanic and Atmospheric Administration. (2010). *Adapting to Climate Change: A Planning Guide for State Coastal Managers*. http://coastalmanagement.noaa.gov/climate/adaptation.html (last accessed 3/20/2014)

National Research Council (NRC). (2009). *Driving and the Built Environment: The Effects of Compact Development on Motorized Travel, Energy Use, and CO2 Emissions, Special Report 298*: Committee for the Study in the Relationships among Development Pattern, Vehicle Miles Traveled, and Energy Consumption.

National Research Council (NRC). (2010). *Informing an Effective Response to Climate Change*. Washington, DC: The National Academies Press.

Newman, P., & Kenworthy, J. (2006). Urban design to reduce automobile dependence. *Opolis: An International Journal of Suburban and Metropolitan Studies, 2*(1), 35–52.

Newman, P., Beatly, T. (2009). *Cities: Responding to Peak Oil and Climate Change*. Washington, DC: Island Press.

Nijkamp, P., & Perrels, A. (1994) *Sustainable Cities in Europe: A Comparative Analysis of Urban Energy Environmental Policies*. London, UK: Earthscan.

Nye Jr., J.S., Zelikow, P.D., & King, D.C. (1997). *Why People Don't Trust Government*. Cambridge, MA: Harvard University Press.

Ockwell, D., Whitmarsh, L., & O'Neill, S. (2009). Reorienting climate change communication for effective mitigation: Forcing people to be green or fostering grass-roots engagement? *Science Communication, 30*(3), 305–327.

Oke, T.R. (1988). Street design and urban canopy layer climate. *Energy and Buildings, 11*(1–3), 103–113.

O'Meara, M. (1999). *Reinventing Cities for People and the Planet*. Washington, DC: Worldwatch Institute.

PEW Center. (2008). *Adaptation Planning: What U.S. States and Localities Are Doing*. Washington, DC: PEW Center on GCC.

Pidgeon, N., & Fischhoff, B. (2011). The role of social and decision sciences in communicating uncertain climate risks. *Nature Climate Change, 1*, 35–41.

Pielke Jr., R., Prins, G., Rayner, S., & Sarewitz, D. (2007). Lifting the taboo an adaption. *Nature, 445*(8), 597–598.

Pitt, D. (2013). Assessing energy use and greenhouse gas emissions savings from compact housing: A small-town case study. *Local Environment: The International Journal of Justice and Sustainability, 18*(8), 904–920.

Pittock, A.B., & Jones, R.N. (2000). Adaptation to what and why? *Environmental Monitoring & Assessment, 61*(1), 9–35.

Pittock, A.B. (2009). *Climate Change: The Science, Impacts and Solutions*. London, UK: Earthscan.

Press, D. (1998). Local environmental policy capacity: A framework for research. *Natural Resources Journal, 38*, 29–52.

Primo, L.H. (1996). Anticipated effects of climate change on commercial pelagic and artisanal coastal fisheries in the Federal States of Micronesia. In J.B. Smith (Eds.), *Adapting to Climate Change: An International Perspective* (pp. 427–436). New York, NY: Spriger-Verlag.

Prinn, R.G., Reilly, J., Sarofim, M., Wang, C., & Felzer, B. (2005). *Effects of Air Pollution Control on Climate*. Cambridge, MT: Joint Program on the Science and Policy of Global Change.

PROVIA. (2013). *PROVIA Guidance on Assessing Vulnerability, Impacts and Adaptation to Climate Change. Consultation document*. Nairobi, Kenya: United Nations Environment Programme.

Ramakrishnan, P.S. (1999). Lessons from the Earth Summit: Protecting and managing biodiversity in the tropics. In M.H.I. Dore & T.D. Mount (Eds.), *Global Environmental Economics* (pp. 240–264). Oxford, UK: Blackwell Publishers.

Read, D., Bostrom, A., Granger Morgan, M., Fischhoff, B., & Smuts, T. (1994). What do people know about global climate change? 2. Survey Studies of educated laypeople. *Risk Analysis, 14*(6), 971–982.

Reichler, T., & Kim, J. (2008). How well do coupled models simulate today's climate? *Bulletin of the American Meteorological Society, 89*(3), 303–311.

Renn, O., & Levine, D. (1991). Credibility and trust in risk communication. In R.E. Kasperson & P.J.M. Stallen (Eds.), *Communicating Risks to the Public* (pp. 175–218). Dordrecht, The Netherlands: Kluwer Academic.

Ribot, J.C., Najam, A., & Watson, G. (1996). Climate variation, vulnerability and sustainable development in the semi-arid tropics. In J.C. Ribot, A.R. Magalhaes, & S.S. Panagides (Eds.), *Climate Variability, Climate Change and Social Vulnerability in the Semi-Arid Tropics* (pp. 13–54). New York, NY: Cambridge University Press.

Rittel, H.W., & Webber, M.M. (1973). Dilemmas in a general theory of planning. *Policy Sciences, 4*, 155–169.

Ruth, M. (2006). *Smart Growth and Climate Change: Regional Development, Infrastructure and Adaption*. Northampton, UK: Edward Elgar Publishing.

Saavedra, C., & Budd, W.W. (2009). Climate change and environmental planning: Working to build community resilience and adaptive capacity in Washington State, USA. *Habitat International. 33*, 246–252.

Saito, I., Ishihara, O., & Katayama, T. (1990). Study of the effect of green areas on the thermal environment in an urban area. *Energy and Buildings, 15*(3–4): 493–498.

Satterthwaite, D. (2008). Cities' contribution to global warming: Notes on the allocation of greenhouse gas emissions. *Environment and Urbanization, 20*, 539–549.

Schaeffer, K. H, & Sclar, E. (1975). *Access for All: Transportation and Urban Growth*. Baltimore, MD: Penguin.

Scheraga, J., & Grambsch, A. (1998). Risks, opportunities, and adaptation to climate change. *Climate Research, 10*, 85–95.

Schneider, S.H., Rosencranz, A., Mastrandrea, M.D., & Kuntz-Duriseti, K. (Eds.). (2010). *Climate Change Science and Policy*. Washington, DC: Island Press.

Short, J.F. (1984). The social fabric at risk: Toward the social transformation of risk analysis. *American Sociological Review, 49*, 711–725.

Simon, H.A. (1956). Rational choice and the structure of the environment. *Psychological Review, 63*, 129–138.

Simon, H.A. (1959). Theories of decision making in economics and behavioral science. *American Economic Review, 49*, 253–283.

Slovic, P., Fischhoff, B., & Lichtestein, S. (1981). Perceived risk: Psychological factors and social implications. In F. Warner & D.H. Slater (Eds.), *The Assessment and Perception of Risk* (pp. 17–34). London, UK: The Royal Society.

Slovic, P., Fischhoff, B., & Lichtenstein, S. (1984). Behavioral decision theory perspectives on risk and safety. *Acta Psychologica, 56*, 183–203.

Slovic, P. (1987). Perception of risk. *Science, 236*(4799), 280–285.

Slovic, P. (1997). Trust, emotions, sex, politics, and sciences: Surveying the risk assessment battlefield. In M.H. Bazerman, D.M. Messick, A.E. Tenbrunsel, & K.A. Wade-Benzoni (Eds.), *Environment, Ethics, and Behavior* (pp. 277–313). San Francisco, CA: New Lexington.

Slovic, P. (2000). *The Perception of Risk*. London, UK: Earthscan.

Somerville, R.C.J, & Hassol, S.J.H. (2011). Communicating the science of climate change. *Physics Today, 64*(10), 48–53.

Smit, B., Burton, I., Klein, R.J.T., & Wandel, J. (1999). The science of adaptation: A framework for assessment. *Mitigation and Adaptation Strategies for Global Change, 4*, 199–213.

Smit, B., Burton, I., Klein, R.J.T., & Wandel, J. (2000). An anatomy of adaptation to climate change and variability. *Climatic Change, 45*, 223–251.

Smit, B., & Wandel, J. (2006). Adaptation, adaptive capacity and vulnerability. *Global Environmental Change, 16*, 282–292.

Steemers, K. (2003). Energy and the city: Density, buildings and transport. *Energy and Buildings, 35*(1), 3–14.

Smith, J. (2005). Dangerous news: Media decision making about climate change risk. *Risk Analysis, 25*(6), 1471–1482.

Spronken-Smith, R.A., & Oke, T.R. (1999). Scale modelling of nocturnal cooling in urban parks. *Boundary-Layer Meteorology, 93*(2), 287–312.

Stern, N. (2008). *The Economics of Climate Change: The Stern Review*. Cambridge, UK: Cambridge University Press.

Tachieva, G. (2010). *Sprawl Repair Manual*. Washington, DC: Island Press.

Tait, M. (2011). Trust and the public interest in the micropolitics of planning practice. *Journal of Planning Education and Research, 31*(2), 157–171.

Thompson, M., Ellis, R., & Wildavsky, A. (1990). *Cultural Theory*. Boulder, CO: Westview Press.

Torvnager, A. (1998). Burdon sharing and adaptation beyond Kyoto: A more systematic approach essential for global climate success. *Environment and Development Economics, 3*(3), 406–409.

Town and Country Planning Association (TCPA). (2007). Climate change adaptation by design: A guide for sustainable communities. http://www.tcpa.org.uk/pages/climate-change-adaptation-by-design.html (accessed 3/20/14)

Tversky, A., & Kahneman, D. (1973). Availability: A heuristic for judging frequency and probability. *Cognitive Psychology, 5*, 207–232.

UKCIP (UK Climate Impacts Programme). (2010). *Identifying adaptation options. London: UK Climate Impacts Program*. http://www.ukcip.org.uk/images/stories/Tools_pdfs/ID_Adapt_options.pdf (accessed 10/04/2014)

UNISDR (United Nations Office for Disaster Risk Reduction). (2012). *Making Cities Resilient Report 2012: My City is Getting Ready! A Global Snapshot of How Local Governments Reduce Disaster Risk, first ed*. UNISDR, for the fifth World Urban Forum.

UN Millenium Project. (2005). *A Home in the City. Task Force on Improving the Lives of Slum Dwellers*. London, UK/Sterling, VA: Earthscan.

Upmanis, H., Eliasson, I., & Lindqvist, S. (1998). The influence of green areas on nocturnal temperatures in a high latitude city (Göteborg, Sweden). *International Journal of Climatology, 18*(6), 681–700.

Van Aalst, M.K., & Helmer, M. (2003). *Preparedness for Climate Change: A Study to Assess the Future Impact of Climatic Changes Upon the Frequency and Severity of Disasters and the Implications for Humanitarian Response and Preparedness*. Hague, Belgium: Red Cross/Red Crescent Centre on Climate Change and Disaster Preparedness.

Van de Vusse, A.C.E. (Eds.). (1993). *Risicocommunicatie: verslag studiedag 17 juni 1993 Wetensc-hapswinkels.* Delft, The Netherlands: Technische Universiteit Delf.

Van Heerden, I., & Bryan, M. (2007). *The storm: What went wrong and why during Hurricane Katrina – the inside story from one Louisianan scientist.* Westminster, UK: The Penguin Group.

Wamsler, C., Brink, E., & Rivera, C. (2013). Planning for climate change in urban areas: From theory to practice. *Journal of Cleaner Production, 50,* 68–81.

Wardekker, J.A. (2004). *Risk Communication on Climate Change,* PhD dissertation, Utrecht, The Netherlands: Utrecht University.

Weber, C., & Puissant, A. (2003). Urbanization pressure and modeling of urban growth: Example of the Tunis Metropolitan Area. *Urban Remote Sensing, 86*(3), 341–352.

Wheeler, S. (2009). California's Climate Change Planning: Policy Innovation and Structural Hurdles. In S. Davoudi, J. Crawford, & A. Mehmood (Eds.), *Planning for Climate Change: Strategies for Mitigation and Adaptation for Spatial Planners* (pp. 125–135). London, UK: Earthscan.

Wilbanks, T.J., & Kates, R.W. (1999). Global change in local places: How scale matters, *Climatic Change, 43,* 601–628.

Wingo, L. (1961). *Transportation and Urban Land.* Washington, DC: Resources for the Future.

Yu, C., & Hien, W.N. (2006). Thermal benefits of city parks. *Energy and Buildings, 38*(2), 105–120.

Zhao, Z.C. (1996). Climatic change and sustainable development in China's semi-arid regions. In J.C. Ribot, A.R. Magalhaes, & S.S. Panagides (Eds.), *Climate Variability, Climate Change and Social Vulnerability in the Semi-Arid Tropics* (pp. 92–108). New York, NY: Cambridge University Press.

3 The global survey on public attitudes towards climate change

Methodology and literature review

There is very little information available on how the public perceives the risks and threats of climate change at local and national levels. Funded by the Foundation of Innovation (Stiftung für Innovation) Rhineland-Palatinate, Germany, and with support by Arizona State University's (ASU) School of Geographical Sciences and Urban Planning as well as ASU's Lightworks initiative, the "Global Survey on Public Attitudes towards Climate Change" research project started in 2010. The research project included a nine-country survey of public perceptions and attitudes in 2010–2011. The nations include the United States, Canada, Mexico, Brazil, Germany, the United Kingdom, the Netherlands, Japan, and Spain. The survey was motivated in part to gain knowledge of the extent to which nations express similar or different viewpoints and perceptions with respect to the various social dimensions of climate change and its impacts. Certainly, we know from the social science literature that threats, whether from natural phenomena and from technological origins are perceived differently based on national experience with the hazard, geography (coastal areas verses inland territory), stage of economic development, and the nature and type of the hazard. There is also evidence that these public perceptions of risk may change over time. Because the idea of global climate change impacts is relatively new, unlike floods and hurricanes, there is little data on how people perceive the causes to be, the nature of the threats over time and space, and our abilities to resolve the problems. We are also unaware of the public's predispositions concerning trust in the information about climate change and their own belief structures. This chapter consists of two main parts. The first part discusses the research methodology utilized in the "Global Survey on Public Attitudes towards Climate Change" research project, especially the survey component.

The second part is a literature review providing an overview of the body of knowledge in the areas of risk perception and communication relevant to the issues of climate change. The literature review is divided into four main sections and first provides an overview of the history and key foundational literature on risk perception and risk communication research. The second section focuses specifically on the public's risk perception of and attitudes towards global climate change. This section is followed by an in-depth discussion of the gap between the science and recommendations provided by the scientific community on the one hand, and the still strong public dissension over climate change on the other. The final section addresses the existing body of knowledge in terms of how existing climate change communication efforts, especially by the mass media, have enforced misconceptions, skepticism, and reluctance to act among the lay public.

Methodology

In addition to various thematic literature reviews, the methods utilized in this research also consist of survey research and analysis at the national and international level to answer the four research questions central to the "Global Survey on Public Attitudes towards Climate Change" research project.

- What are the public's perceptions of climate change in terms of threat and risk, saliency of the issue, trust in climate change information, and acceptable public strategies?
- What importance do climate change risk perceptions and attitudes play in the public's willingness to support mitigation and adaptation strategies?
- How do the public perceptions regarding climate change and attitudes towards mitigation and adaptation strategies vary by socioeconomic factors?
- What role do levels of knowledge and perceptions of trust and responsibility play in the public's level of support for adaptation and mitigation policies?

Based on the data collected, one of the main goals of this long-term research project is to identify common themes among the populations of different countries in terms of climate change risk perception, attitudes, and behavior that can inform and improve communication efforts on the international scale. Furthermore, the project also points to differences between countries related to the public perceptions of climate change and its inherent issues. These country-specific characteristics need to be acknowledged and addressed by communication efforts at the national scale.

Analytical framework

The following describes the project's analytical framework for the research as shown in Figure 3.1. The study is divided into four different phases. Phase 1 included a comprehensive literature review, which provided the foundation for the research objectives and the rationale for risk perception research on climate change, and it certainly informed us of the importance of working with the key variables and questions during the development of the survey instrument. Phase 1 also included the development and the survey instrument. Phase 2 consists of the data collection by launching the survey in different countries. The survey instrument and the process of the data collection are discussed in detail in the following sections. After all data are collected, Phase 3 focuses on

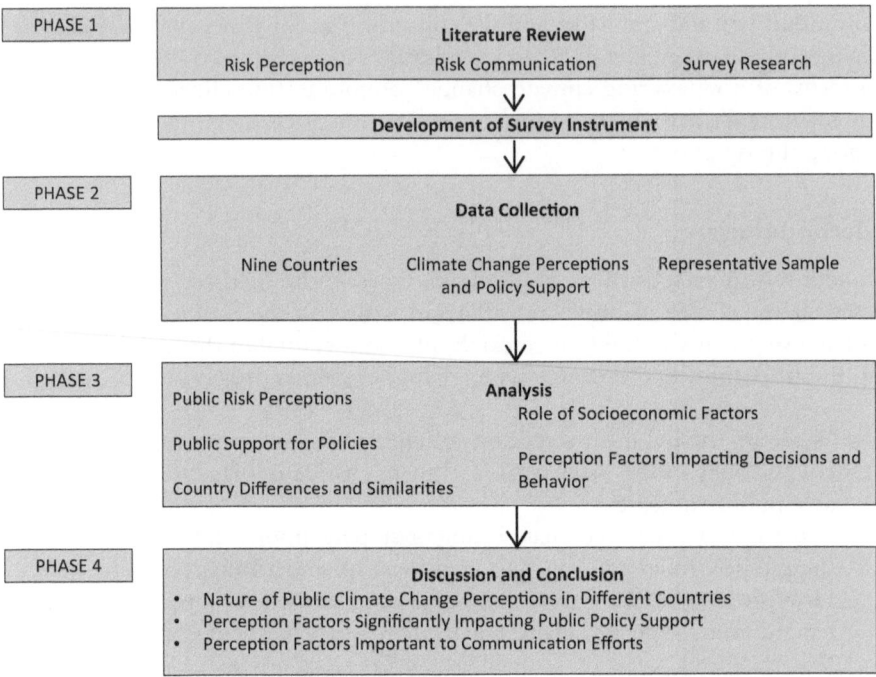

Figure 3.1 Analytical framework for the "Global Survey on Public Attitudes towards Climate Change" project

Source: Author's illustration.

the analysis of the survey data guided by the four research questions listed in the previous paragraph. The statistical methods applied consist of frequency analysis, cross tabulation, as well as different regressions and are discussed in depth together with the underlying hypotheses in a later section of this chapter. The final phase consists of the discussion and conclusions focusing on the goals of this research endeavor; identifying the factors which help to understand the nature of public perceptions of global climate change in different countries as well as to identify perception factors which have a significant impact on the public's level of support for mitigation and adaptation policies, the willingness to commit to behavioral changes, and on the success of future communication efforts. Moreover, the newly gained knowledge is put into perspective with the existing body of literature and the underlying theories discussed in Chapter 1.

Survey research

Phase 2 starts with the implementation of the international survey, followed by the analysis of the retrieved date in Phase 3. Survey instruments present a valuable tool to capture the public's perceptions towards global climate change and, as illustrated in Figure 3.2, is a key component of this study. Currently data from nine countries are available and being analyzed using internet panels, ensuring a demographically representative sample for each country. This allows sampling a large population, while asking the same questions, thus establishing consistency and collecting standardized, quantifiable, and empirical comparative data. The internet panels present a cross-section of all age groups of 18 years and above, gender, income groups, different regions of the country, and level of education. Relying on internet panels as sampling frames has several advantages over other survey methods, such as telephone, mail, or personally administered surveys (Fowler Jr., 2008). Email or web-based surveys allow coverage of a wide geographical area with relatively low costs. In addition, internet surveys allow the participants to choose their own time to answer the questions, which can increase the response rates (Babbie, 2007). Furthermore, the responding panel members know that they will remain anonymous throughout the entire process. It is impossible to link certain answers to certain people since the database only assigns an ID number to each person without their name or address. This is very important, since the survey addresses personal feelings, behavior, and knowledge regarding the issue of climate change.

The surveys' household selections were random within the parameters of socioeconomic categories and ownership of computers. The total sample size accounted for 7,327 households. Each country's sample size was large enough to provide the ability to generalize to each country with a 95 percent confidence level and a ± 4 percent margin of error. The sample size per country ranges from 539 for Canada to 947 participants in the United States. Figure 3.3 shows the spatial distribution of participants for each of the countries.

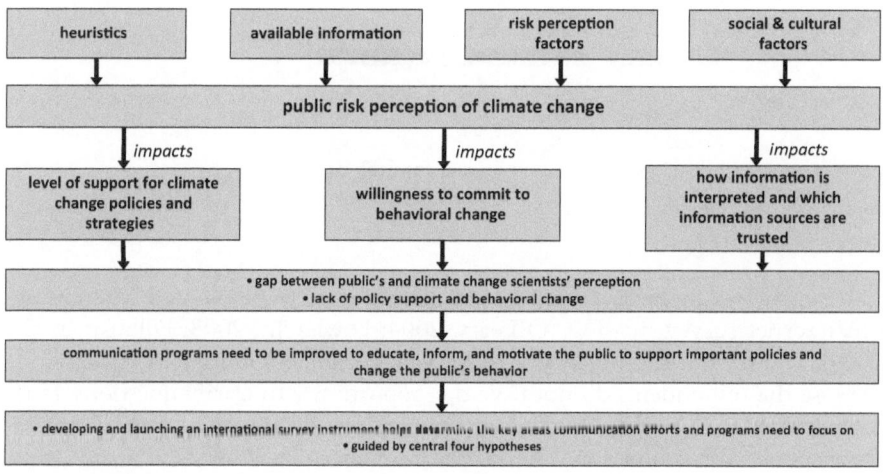

Figure 3.2 Risk perception and survey research framework

Source: Author's illustration.

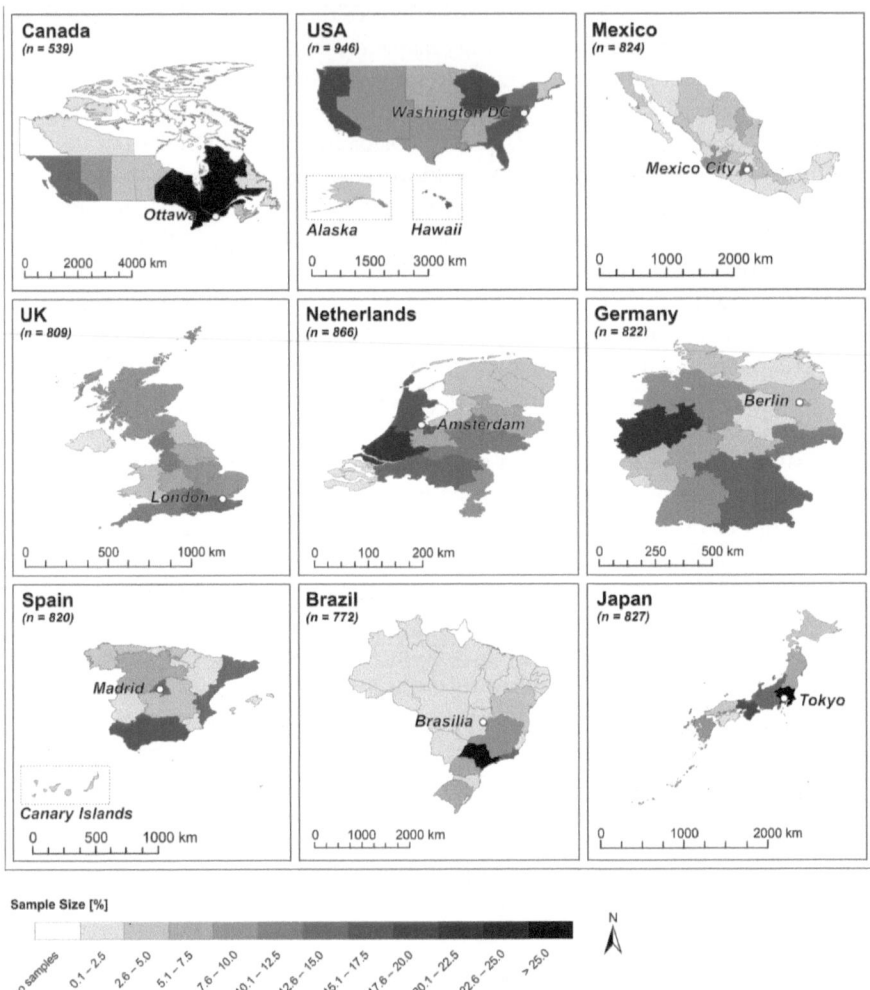

Figure 3.3 Spatial distribution and sample sizes for each country
Source: Author's illustration.

Nevertheless, there are also some pitfalls that need to be avoided when using this internet survey method (O'Leary, 2004; Fowler Jr., 2008; Dillman et al., 2009). The survey questions have to be well designed and easy to understand, because the respondents do not have the opportunity to clarify questions. Furthermore, it is very likely that follow-up emails need to be sent out; otherwise the response rate can be low. The survey instrument itself needs to avoid complex terms and language or double negatives. Both can confuse the participants of the survey. Other pitfalls are created by poorly wording are ambiguous and double-barreled questions. Ambiguous questions can happen easily, because frames of reference can be highly divergent. Double-barreled questions address more than one issue, but only ask for one response. In both cases, the answer cannot be analyzed beyond doubt. Biased, leading, or loaded questions and statements, present another common pitfall in survey research. So-called ring true statements are phrases people are likely to agree with because of the tone.

There is also the danger of placing statements in surveys to which in general the respondent agrees but not without elaborating. Instead, the respondent is forced to either agree or disagree. Formulating leading questions can happen quite easily and is often done for political purposes. However, questions and statements that are not unbiased or specific will lead to viable results. Other aspects being problematic for the respondent are recall dependent questions, offensive questions, questions that assume certain knowledge, and questions with unwarranted assumptions (O'Leary, 2004; Fowler Jr., 2008). The questions should be relevant, and the respondents must be willing to answer to generate trustworthy data leading to credible results and conclusions.

Survey instrument

Based on the theoretical framework, the survey covered a wide range of questions relevant for analyzing the public's climate change risk perceptions and attitudes, their willingness to support climate policies, and their readiness to commit to behavioral changes. Questions were divided into different thematic groups. The first set of questions focused on the importance the public placed on the government to reduce, prevent, or improve upon various societal problems. Questions in this section provide data necessary to determine how important climate change is to the public in the context of other societal challenges, or the political saliency of the climate change problem.

The second group of questions captured the participants' level of awareness and knowledge of various climate change aspects. For example, questions focused on what climate change means and whether or not the participants feel informed in terms of climate change causes, impacts, and existing mitigation or adaptation measures. The third group of questions explicitly addressed the public's perceptions about climate change risks and threats. Participants answered questions about their level of concern regarding various possible climate change impacts. Other questions presented participants with scientific facts and other climate change statements to determine the public's attitudes and beliefs regarding existing scientific data, the scientific community, causes of climate change, support for renewable energy, and other perceptual and attitudinal factors.

Proven to be a critical factor in risk perceptions, especially for hazards with high uncertainty, the survey instrument also addressed the importance of trust. The role of trust is an important perceptual dimension that influences the success of climate change policies as well as the public's willingness to commit to behavioral changes. As suggested in the literature (Read et al., 1994), the public is more likely to support climate change policies if they trust the science behind it and the source of information. Furthermore, research shows that the failures of risk communication are significantly influenced by the public's trust in the communicator and in the ability of certain individuals, industries, or institutions responsible for risk management (Kasperson et al., 1992; Nye Jr. et al., 1997). Pijawka and Mushkatel (1991) found that low trust factors in the technology and regulatory institutions were associated with high-risk perceptions and public concerns over the proposed high-level waste repository. Public trust in organizations whose risk management addresses adaptation and mitigation strategies is vital in order to generate social cooperation and to increase the likelihood of successful resiliency policies.

Questions were closed-ended, multiple-choice questions with Likert-type scaling, which were balanced equally. To prevent biases, balanced Likert scales give equal weight to the number of favorable and unfavorable answer categories. The Likert scales used in the survey instrument were mostly 5- to 7-point scales with many of the answers ranging from strongly disagree to strongly agree with a neutral answer option in-between. The survey instrument was tested and reviewed by social science researchers experienced in survey research to ensure its validity. The questionnaire was also tested and reviewed to validate if the English version would correspond to other languages in terminology and meaning.

Criteria for selection of surveyed countries

To ensure credibility, the countries selected for this study must establish validity or authenticity within the surveyed population sample (Yin, 1994). This means that the findings can directly be assumed to a larger population. The use of internet panels requires that people from all social groups must have access to the internet and the skills necessary to participate. In the case of Africa, this cannot be guaranteed and thus no countries representing the African continent are surveyed in the "Global Survey on Public Attitudes towards Climate Change" research project. The internet is not as available in rural developing areas as it is in developed countries and the limited amount of time and funding disallows face-to-face surveys for these countries. Therefore, the countries participating in this study are the United States, Canada, Mexico, Brazil, Spain, Germany, United Kingdom, Netherlands, and Japan.

The United States is an obvious choice as it (a) is a Superpower, (b) is the richest country in the world, and (c) has the second highest greenhouse gas (GHG) emissions (World Bank, 2014a). In addition, there have been significant national political debates over climate change. Global agreements to reduce GHG emissions and adapt to climate change stand and fall with the involvement of the United States. Furthermore, major natural disasters in the United States have been linked to climate change at least in the media. Finally, the little that is known about climate change perceptions research is dominated by the United States. Therefore, the results and findings of this paper can be compared to earlier studies providing research validation for some questions. Mexico is among the top 20 largest economies in the world, and former president Calderon ranked the environment and climate change very high on his priority list (Booth, 2010). However, the country's goal can only be achieved if the public supports climate change policies and measurements. Moreover, without public support the newly elected president Enrique Peña Nieto may be hard pressed to achieve the goal to cut GHG emissions by 30 percent from business-as-usual levels by 2020 and by 50 percent by 2050 set by the past administration (Teixera, 2012). Despite its level of economic development, it (a) is still characterized as a developing country, (b) has experienced coastal storms, and (c) has a strong policy interest in sustainability practices. Canada's annual temperature increased 1.2°C between 1948 and 2005. This increase is significantly higher than the global average of 0.74°C (IPPC, 2007). As a result, sea level rise could become a serious threat to coastal communities requiring adaptive measurements supported by the public. Moreover, three of Canada's major cities are ranked in the top ten cities of the world on the urban resiliency rating system (Barkham, 2014).

Germany is located in the center of the European Union (EU) and is among the Group of Eight (G8) highly industrialized nations with one of the strongest economies worldwide. Within the EU, Germany is a driving force for achieving a global climate treaty (Weidner & Mez, 2008). Furthermore, Germany was less affected compared to the rest of the EU during the current economic downturn. As a result, the country will most likely have even larger political capital in the near future, impacting European climate change policy (Hill, 2011). Germany has taken a leadership role when it comes to climate change. Spain, on the other hand, is among the countries in Europe heavily impacted by the worldwide economic crisis. The country has already decreased its investments in renewable energy significantly to offset these impacts (Pew Research Center, 2010). Nevertheless, its extensive shoreline and the potential for tourism make successful adaptation to sea level rise and possible extreme flooding events key to Spain's future. The Netherlands is widely considered a trendsetter when it comes to successful planning policies and strategies. This is evidenced by the country's ability to successfully cope with weather extremes such as storm floods and sea level rise. With one-third of the country below sea level, however, its strategies are bound to be tested when levels rise as anticipated due to climate change. The fourth European country surveyed in this study is the United Kingdom. Together with Germany and France, the United Kingdom is a very influential country within the European Union and worldwide. In terms of climate change, its national government launched a campaign in 1991 for energy conservation with the goal to educate the public about the global implications of local actions (Hinchliffe, 1996).

Japan was chosen because it is an insular state and a driving economic force of the region. It is also part of the world's largest and most populous continent, Asia. Furthermore, one of the major impacts of climate change that it will experience will be ocean-related disasters. Brazil plays a leading role in South

America with the largest economy of the continent (World Bank, 2014b). The country is also home to the Amazon, one of the largest ecosystems on the planet. Its forests are extremely vulnerable to climate change impacts (Malhi et al., 2008). Certainly the study can be expanded to other countries and plans are ongoing to do exactly that and to conduct repeat studies over time to see longitudinal changes in perceptions and the reasons for these.

Statistical analysis

The data gathered from the survey are mostly analyzed using a wide range of statistical methods. This "Global Survey on Public Attitudes towards Climate Change" research project applies basic statistical methods such as frequency distributions and descriptive statistics, as well as advanced methods including standard multiple regression, stepwise regression, and one-way independent Analysis of Variance (ANOVA; Field, 2013). In order to compare the means of several survey answers between countries, the standard deviations were calculated to ensure that the mean is a good representation of the data. The Kruskal-Wallis test was performed to examine if the means are also significantly different between the countries. The Kruskal-Wallis test is the non-parametric counterpart to ANOVA analysis. It enables the comparison of the means of multiple populations. The survey data collected were not normally distributed nor could the homogeneity of variance be assumed. The Kruskal-Wallis test was used as it is free of assumptions about how data are distributed and does not require homogeneity of variance to test if means are significantly different between groups. The output of this test in SPSS predictive analytics software includes a significance value. As long as this value is below .05, the country means are significantly different. The Kruskal-Wallis test also provided mean rank scores, which groups as well as identifies outliers among the nine countries. All regression results were considered significant at $p < 0.001$, which means that there is less than a 0.1 percent chance that the particular F-ratio would happen if the null hypothesis were true.

In addition, based on the collected survey data, new variables were coded and indexes were created to gain further insight and a more holistic understanding of the interrelationships of risk perception and risk communication as well as the barriers to successful public climate change communication. For example, the survey instrument does not directly ask if the participants belong to one of the three groups: those who do not believe climate change is happening at all, those who think climate change is happening but is a natural event and not human caused, those that believe it is happening and is human caused. However, answers to different survey questions were used to determine to what degree the participant believes climate change is real and allows the creation of a new Climate Change Believer variable based on these specific answers. This new variable coded allows categorizing the survey participants into different groups based on their attitudes towards the reality of climate change and if it is natural or human caused. These types of additionally created variables were used as independent and dependent variables in different regressions in order to analyze the research questions and underlying hypotheses.

Within the scientific literature there is an ongoing debate about using Likert-type data and scales for standard multiple regression analysis (Brown, 2011; Jamison, 2004). The debate focuses on the question if Likert-scales can be treated as interval data, which is a key assumption that has to be met for multiple regression analysis (Field, 2013). Skeptics argue that data are lost if Likert scales are treated as interval data resulting in underestimating the actual strength of the relationship (correlation coefficient [R]) between the predictor ant outcome variables (Owuor, 2001). However, a study by Labovitz (1975) tested the differences between using ordinal categorized data and continuous variables in regression analyses and concluded that categorical data, such as Likert-type scaling, can be analyzed as continuous data. This finding was further supported by Jaccard and Wan (1996) whose statistical tests show that not using "true" interval data does not greatly affect Type 1 or Type 2 errors. This means that is highly unlikely that standard regressions based on Likert-type data would show false relationships between variables that do not exist in reality. In addition, different quantitative studies in the field of medical and psychology research (Baggaley & Hull, 1983; Maurer & Pierce, 1998, Vickers, 1999) proved that Likert-scales can indeed be analyzed effectively as interval scales and fulfill all the assumptions needed for the standard and stepwise regression methods applied in this research study.

Despite the ongoing discussion among scientists and statisticians, especially in the field of social sciences, in which this research is situated as well, Likert-type data are consistently treated as interval data and used for regression analysis (Johnson & Slovic, 1995; Peters et al.. 1997; Sjoeberg, 1998; Leiserowitz, 2006). For example, in their published study "Presenting Uncertainty in Health Risk Assessment: Initial Studies of Its Effect on Risk Perception and Trust". Branden B. Johnson and Paul Slovic (1995) showed through multiple regression analysis based on Likert-type survey questions that public reactions to environmental problems are less impacted by the presentation in the media compared to general factors of risk attitudes and perceptions. Peters et al. (1997) focused on the role of the perception factors trust and credibility as key elements in environmental risk communication. The study applied standard multiple regressions to test the hypothesis that trust and credibility are strongly impacted by three independent factors – perceptions of knowledge and expertise, perceptions of openness and honesty, and perceptions of concern. The majority of these factors were measured through survey questions using four-point, Likert-type scaling. Based on the psychometric paradigm, which is also the methodological foundation of this research, Sjoeberg (1998) examined the relationships between worldviews, political attitudes, and risk perceptions using multiple regressions with scores provided by Likert-type data. The study showed that approximately 10 percent of the variance in one factor could be explained by the remaining two. Another study, specifically in the context of climate change, used survey data from the United States used multiple regressions to test if climate change risk perceptions and policy support are influenced by experiential factors (Leiserowitz, 2006). The details of this study are discussed in the second half of this chapter, but the study did use a similar analytical framework as the research conducted as part of the "Global Survey on Public Attitudes towards Climate Change" research project. These studies are well known and their findings are considered as important contributions to the existing body of knowledge.

Nevertheless, additional steps were taken to further decrease the likelihood of information loss and wrong results as well as to acknowledge the arguments by skeptics, who caution the use of Likert-type scales as interval data. Research suggests that when Likert-type data are used is multiple regression analysis the estimates improve if the answer scales have more than three points and a sample size of 300 participants (Owuar, 2001). Both points are considered "Global Survey on Public Attitudes towards Climate Change" research project, as no Likert-scale used has less than 4 points and the smallest country sample consists of over 500 people. Furthermore, Brown (2011) argues that indexes created from Likert-type data not only further reduce the likelihood for errors (Jaccard & Wan, 1996; Jamison 2004), but are actually "true" interval data. As a result, the vast majority of regressions performed in our research use different additive indexes from Likert-type survey questions as dependent variables.

The additive indexes created from the survey data focus on the areas of the public's general support for mitigation and adaptation policies in general, their preparedness to change their behavior, as well as their willingness to pay more for climate change policies. Furthermore, indexes are also created for specific themes. For example, in addition to the general indexes for the public's support for mitigation or adaption strategies, indexes are also created for the public support regarding energy efficiency policies, economic incentives, and planning strategies. These different indexes are created based on different sub-questions, which have in common an overarching theme such as general support for mitigation polices or general willingness to commit to behavioral changes. Similar as the new variables mentioned above, the different indexes will be used in various regressions as dependent and independent variables. The main indexes created for this study are the following:

- Index 1: Overall support for mitigation polices
- Index 2: Overall support for adaption polices
- Index 3: Overall public willingness to pay more for climate change strategies
- Index 4: Overall public willingness to commit to behavioral changes
- Index 5: Overall level of consequences perceived by the public from environmental changes
- Index 6: Overall public's perceived level of threat from climate change

The basic analysis consists of frequencies and various cross-tabs for all nine countries combined and for each country separate. The basic analysis of the frequencies and percentages of the survey answers allowed comparing the countries and identifying the significant differences between countries in various areas that are addressed in the survey instrument. Together with the results from various crosstabs between indexes, differences, and possible correlations identified were explored through advanced statistical methods such as different regressions and ANOVA. The descriptive analysis also prepared the large amount of survey data into a manageable size. Standard multiple regressions were used to illustrate how independent variables (such as demographics, attitudes towards climate change, or trust in climate science) are related to the dependent variables (such as willingness to pay, policy support, or person connection towards climate change). These relationships were further explored through stepwise regressions to determine the subset of independent variables that has the strongest relationship to each dependent variable. It is important to note, that throughout the statistical analysis, depending on the underlying hypothesis, numerous variables were used as independent as well as dependent variables in different regressions. Since multiple independent variables per regression can raise the R^2 and be a potential source for error, this study reports on the adjusted R^2. The adjusted R^2 compensates for the use of more predictors and adjusts the value downwards (Field, 2013). The results from the survey research and from the previously conducted literature review provided the insights necessary to address the research questions of "Global Survey on Public Attitudes towards Climate Change" research project. Furthermore, this analysis points out possible misconceptions among the lay public, identify trusted communication channels, and identify the key areas communication efforts need to focus on to improve the success of current and future climate change policies.

Underlying hypotheses and main variables

The following section provides an overview and justification of the main hypotheses, variables, and statistical methods used to address the research questions during the fourth phase of the "Global Survey on Public Attitudes towards Climate Change" research project.

Hypothesis 1

"The public's perceptions of climate change in terms of threat and risk, saliency of the issue, trust in climate change information, and acceptable public strategies vary among countries".

This research is based on the psychometric paradigm (Fischhoff et al., 1978; Slovic et al., 1984) that implies that risk means different things for different people. Therefore, we can assume that public perceptions in the context of climate change vary among countries with different cultural and economic backgrounds. This hypothesis was tested by comparing the basic frequencies and means of the relevant survey questions and determining statistically differences among the answers provided by the surveyed population within the nine countries. In addition to these questions the results of four different indexes were compared as well. The indexes present the overall support for mitigation and adaptation policies, the public's general readiness to change their behavior, as well as their willingness to pay more for climate change strategies. This basic analysis relying on descriptive statistics helped identifying patterns between countries and provided the foundation for the more complex statistical analytical methods, which were necessary for the following hypotheses.

Hypothesis 2

"The public's general support for mitigation and adaptation policies is linked to the way they perceive (1) the level of consequences from possible environmental changes and (2) the general level of threat resulting from climate change". Research shows that risk perception has a significant impact on individual and group behavior and thus needs to be considered when developing global climate change policies and strategies (Slovic, 2000). For example, research linking perceptions of risk to acceptance and opposition to specific technologies, such as nuclear power have shown that the higher the perceived risk the higher the opposition by the public towards such technologies (Slovic et al., 1981). Furthermore, recent studies emphasize the important role of emotions, such as level of concern and perceived level of threat, in the decision-making process (Finucane et al., 2000; Loewenstein et al., 2001; Paton, 2008; National Research Council [NRC], 2010a). This leads to the assumption that such relationships also exist in the context of natural hazards resulting from global climate change. Therefore the hypothesis was developed to test if the public's support of policies and strategies to reduce the causes and impacts of natural hazards resulting from climate change also correlate to the public's risk perception of climate change?

The hypothesis was tested through frequency distribution analysis and crosstabulations. The frequency analysis focused on survey questions addressing the public's perceived level of consequences from environmental changes over the next 20 years, the level of support for mitigation and adaptation policies, and the perceived level of threat resulting from climate change over the next 50 years. In a second analytical step crosstabs were created to test for relationships between mitigation and adaptation attitudes (support) and the public's perceived level of consequences from different environmental changes such as global climate change and level of threat resulting from global climate change in particular. The crosstabs were used to identify differences among the nine countries in terms of the significance of these relationships.

Hypothesis 3

"The public's perception towards climate change is the main reason for the (1) low policy support, (2) willingness to pay for climate change policies, and (3) willingness to change their behavior related to mitigation and adaptation". As pointed out by the literature (Rotter, 1966), if the public believes that they can control the events that affect them, they also are likely to feel that applicable strategies exist in which they can be engaged in (locus of control concept). However, climate change is characterized by high uncertainties, unfamiliar risks, and other characteristics of hazards which make a personal connection and engagement more difficult (Maibach et al, 2009; O'Neil & Hulme, 2009; Whitmarsh et al., 2009). Research suggests that current public engagement with global climate change is low (Leiserowitz, 2005) impacting public policy support and behavior.

The hypothesis was tested by conducting various multiple regressions, using six independent variables and sixteen dependent variables. The public's preference towards four general climate change strategies and the level of belief in the reality of global climate change were used to describe a person's attitude towards climate change. The remaining four independent variables focused on the public's level of concern regarding possible dangerous impacts of climate change on different geographical and personal levels as well as time-scales. The dependent variables used for the regressions included different additive indexes, based on survey answers from various survey questions, as well as single survey questions addressing the public's willingness to pay more or commit to behavioral changes to mitigate and adapt to climate change.

Hypothesis 4

"The general attitude and public risk perceptions of climate change can be largely explained by socioeconomic variables". This is another key main hypothesis that needs to be tested. Similar to the previous hypothesis two main regression analyses types are utilized consisting of standard multiple regressions and stepwise regressions. The independent variables include various socioeconomic characteristics of the surveyed population such as age, gender, household income, and level of education. The dependent variables cover the answers to different survey questions related to public's climate change perception in terms of its risks, threats, level of concern and consequences, timeframe for potential impacts, and behavioral attitudes towards specific mitigation and adaptation policies.

The reason for testing the impact of different socioeconomic characters of the survey participants on general climate change attitude and public risk perceptions is rooted in past perception research in the areas of technological and natural hazards (Burton et al., 1978; Douglas & Wildavsky, 1982; Short, 1984; Slovic, 2000) arguing that perceptions are socially constructed and can vary by culture, human development, affluence, and demographics. Furthermore, this research is a multinational study including different socioeconomic characteristics. As a result, we can assume that socioeconomic characteristics are a potential factor impacting public's risk perception and policy support.

Public risk perception and communication

Since the beginning of the 1980s, the body of knowledge has grown considerably in the fields of risk perception and risk communication (Slovic et al., 1981; Slovic, 1987, 2000; Wardekker, 2004). A major reason for this increase in knowledge was due to industrialists and regulators who recognized that the public believes that they are faced with more risk today than in the past and that levels of risk will continue to increase in the future (Slovic, 1997). Studies done in the fields of geography, sociology, political science, anthropology, and psychology contributed significantly to the current understanding of risk perception (Slovic, 2000).

Whereas geographical research first focused on examining the human behavior faced by natural hazards, studies addressing risk perception and behavior later included technological hazards as well (Burton et al., 1978). Relevant literature (Kasperson et al., 1995; Krimsky & Golding, 1992; Slovic, 2010) shows that threats from natural phenomena and technological origins are perceived differently based on the nature and type of hazard, experience with the hazard, geography (coastal areas versus inland territory), and/or stage of economic development. There is also evidence that public risk perceptions may change over time (Loewenstein & Mather, 1990; Gomez et al., 1992; Tate et al., 2003). Risk perceptions also vary by the nature of the hazard in terms of hazard consequences, level of uncertainty, whether the hazard is voluntary, known or unknown, and dread, among other heuristics. Moreover, sociologists (Short, 1984) and anthropologists (Douglas & Wildavsky, 1982) discovered that that perceptions and risk acceptance are embedded in social and cultural contexts and that friends, family, coworkers, and respected public officials are responsible for transmitting many of the social influences affecting individual's response to hazards.

The field has since added work on technological hazards as well (Bowonder et al., 1985; Kates et al., 1985; Flynn et al., 1995). Research within the field of psychology started with empirical studies addressing probability assessment, utility assessment, and decision-making processes (Edwards, 1961). In psychology, the work by Paul Slovic (2000) has illuminated the field of human response to both natural and technological hazards and disasters. Some influential findings on hazard perceptions came out of the interdisciplinary research on siting the high-level nuclear waste repository that includes consideration of the social amplification of risk concept, public trust factors, and the role of scientific uncertainty (Kasperson et al., 1988). Recent studies also emphasize the important role of emotions in the decision-making process (NRC 2010b; Paton 2008).

Risk communication first focused on public misconceptions regarding risk (Wardekker, 2004). Initially not much attention was paid towards the public's perceptions, instead only the expert's estimates were acknowledged (Department of Health [UK], 1997). During this time, the typical method for risk communication was to "put risk into perspective", and long lists of numerous risk comparisons were created. These comparisons, however, can be misleading, dissatisfying, and are difficult to use responsibly and effectively. Therefore, such comparative lists can be counterproductive (Freudenberg & Rursch, 1994) and even threaten the credibility of the risk communicator (Slovic, 2000). In time, researchers started to acknowledge the importance of lay risk perception and studies began to examine what actually does cause public concern and why. Since the 1980s, the growing body of knowledge shows that perceived risk is both quantifiable and predictable (Slovic, 1987; Gaerling & Golledge, 1993) and that psycho-metric techniques are applicable to identify similarities as well as differences regarding risk perceptions and attitudes among groups (Slovic, 2000; Simmons, 2007). Moreover, research shows that the concept of risk means different things to different people.

Risk communication efforts are often needed to present and simplify complex technical material, influenced by uncertainty, and difficult for laypersons to understand (Slovic, 1986). Thus, in order to enhance successful programs, communicators must gain a good understanding of the limitations of current scientific risk assessment and the idiosyncrasies of the human mind. In particular, it is important to realize that the public's risk perceptions are rooted in theoretical models based on assumptions and subjective judgments. Thus, incomplete assumptions and judgments most likely result in inaccurate risk assessments or perceptions. Furthermore, even when faced with solid evidence through information and educational programs, research shows that people's beliefs change slowly and disagreements about risk do not automatically disappear (Nisbett & Ross, 1980). Strong initial views are resistant to change, because once they are formed, they influence how subsequent information is interpreted. Thus, new evidence appears reliable and informative if it is consistent with one's initial beliefs; contrary evidence on the other hand, is likely to be viewed as unreliable, erroneous, or unrepresentative (Wardekker, 2004).

However, if people do not have strong prior opinions the situation is quite different. In that case, these persons are deeply influenced by the way the risk is formulated and presented to them (Tversky & Kahneman, 1981). The fact that subtle differences in how risks are presented can have marked effects suggests that those responsible for information programs have considerable ability to manipulate perceptions and behavior. This possibility raises ethical problems that must be addressed by any responsible risk-information program (Wardekker, 2004).

Taken as a whole, insights gained by researching public risk perception have important implications for communication efforts (Slovic, 1987). The public's basic conceptualization of risk does reveal important concerns to communicators, which experts tend to overlook in their risk assessment (Renn & Levine, 1991). Relying on statistics alone is not enough for guiding personal or public decision policies. Instead, risk perception is not only determined by accident probabilities, annual mortality rates, or mean loses of life expectancy, but also by numerous other characteristics of hazards such as uncertainty, controllability, catastrophic potential, equity, and threat to future generations (Wardekker, 2004). The classic risk perception factors (Slovic et al., 1981; Slovic, 1987) can be organized into different dimensions and are as follows:

- Dread Risk

    o Controllability; Dread; Global/not-global catastrophic; Fatal/not-fatal consequences; Equity; Catastrophic/individual high/low risk to future generations; Easily/not-easily reduced; Risk increasing/decreasing; Voluntary/involuntary; Clarity and importance of expected benefits; Harmful intentionality; Inescapable by taking personal precautions; Manmade rather than natural sources

- Unknown Risk

  o Observable/not-observable; Known/unknown to those exposed; Effects immediate/delayed; Old/new risk; Risks known/unknown to science; Contradictory statements from responsible sources

- Exposure

  o Number of people exposed to the risk; Personal exposure; Identifiable rather than anonymous victims

- Other

  o Probability of undesired consequences

Research suggests that risk perception factors belonging to the area of Dread Risk have the biggest impact on what risks are considered as high, why people want a risk reduced, and why they call for strict regulation (Slovic, 1987). For example, the location of nuclear facilities falls into this category, whereas climate change can be considered to most people as an unknown risk.

Public perceptions of climate change

The existing body of knowledge in the field of public perception regarding climate change has been growing considerably (Jin & Shriar, 2013; Leiserowitz, 2005, 2010; Whitemarsh et al., 2011; Fischer et al., 2012). Research shows that risk perception has a significant impact on individual and group behavior and thus needs to be considered when developing climate change policies and strategies (Kahan, 2012; Akerlof et al., 2013). Nevertheless, climate change perception is still a relatively new survey topic. The existing body of literature also suggests that perceptions change over time due to factors such as extreme events, amount of media coverage or level of reporting, economic conditions, scientific information, values and worldviews, among other factors.

In contrast to the very limited number of studies on climate change perceptions at the international level, there is a significant body of knowledge available for the United States. Although no formal national assessment exists in terms of the public perception of climate change, numerous representative scientific studies and different opinion polls do provide important insights. The first surveys in the United States were conducted in the early 1980s, but strong public interest did not emerge before 1988 (Bord et al., 1998). The key year was 1988 for the development of public concern for climate change for two reasons. First, at that time, the United States was hit by a severe drought and heat wave. Second, and even more important, James Hansen, who at the time was the director of National Aeronautics and Space Administration's (NASA) Goddard Institute for Space Studies, gave testimony before Congress that global climate change had begun. As a result, more empirical studies were conducted, and media coverage as well as public interest increased in the United States.

Overall, the existing body of climate change research and the perception surveys supports the argument that the American public is aware of global climate change, believes that it is real and is highly concerned about it (Bostrom et al., 1994; Read et al., 1994; Bord et al., 1998; Henry, 2000; Seacrast et al., 2000; Department for Environment, Food and Rural Affairs [DEFRA], 2002, 2007; Dessai et al., 2004; Leiserowitz, 2005; Ockwell et al., 2009). However, research also identified contradictions in American climate change risk perception and policy preferences (Rosenstone et al., 1997; O'Connor, 1999; Moser, 2006; Leiserowitz et al., 2010). On the one hand, the US public strongly supports a range of national and international policies to mitigate climate change; on the other hand, several carbon tax proposals are strongly opposed. Thus, the public indeed largely supports policy action at the national and international scale, but resists tax policies that directly affect them. At this juncture, very little is known about the level of public acceptance and willingness to support climate change policies and possible international differences. However, having such knowledge would point out key areas communication programs would need in order to increase public support for climate change policies developed in their own country and internationally. Therefore the research from the "Global Survey on Public Attitudes towards Climate Change" research project presented in Chapters 4 and 5 test public support and acceptance for a set of adaptation and mitigation strategies in different countries as well as their willingness to commit to behavioral changes.

Furthermore, existing data suggest that that current public engagement with climate change is low. Despite the clear implication of mitigation strategies for individual values, choices, and behaviors the demand for energy for domestic use and transportation is increasing in many developed countries. Moreover, pro-environmental behavior as a response to the causes and possible negative impacts are even more limited (Maibach et al., 2009; O'Neil & Hulme, 2009; Whitmarsh, 2009). We now know that only a limited number of people are willing to do more than advance domestic energy conservation and even less are prepared to take actions to adapt to climate change.

Risk perception literature draws from the concept *locus of control*, which refers to the extent to which individuals believe that they control events that affect them (Rotter, 1966). For many hazards the public feels that applicable strategies exist in which they can be engaged in. However, climate change is characterized by high uncertainties, unfamiliar risks, and other characteristics of hazards which make personal connections and engagement difficult. This if further emphasized by the fact that climate change is more and more considered as a "Black Swan" (Curry, 2011a; Taleb, 2010; Winston, 2010). Black Swan rare events are characterized by high uncertainties, are unanticipated, and lead to misconceptions among the lay public. The theory was developed to explain

> 1) the disproportionate role of high impact, hard to predict, and rare events that are beyond the realm of normal expectations in history, science, finance and technology, 2) the non-computability of the probability of the consequential rare events using scientific methods, 3) the psychological biases that make people individually and collectively blind to uncertainty and unaware of the massive role of the rare event in historical affairs.
>
> (Curry, 2011b)

In fact, the argument can be made that the issue of climate change includes two Black Swans (Winston, 2010). The first Black Swan is climate change itself with all its uncertainties, misconceptions, and the clear gab between the recommendations provided by the scientific community and public attitude and behavior. The second Black Swan is the global effort necessary to successfully mitigate and adapt to climate change. The policies needed and behavioral changes required to reduce climate change will require a fundamental shift away from today's business, policies, and lifestyle models. If and how the necessary policies will be ever developed and implemented is very uncertain, especially since past global climate change treaties have mostly failed. Many hope that effective communication efforts can foster a personal connection to climate change, raise the level of concern, and thus increase the level of support for mitigation and adaptation policies as well as the willingness among the public to engage in a more sustainable behavior.

In the case of the United States, existing surveys show that although concern about climate change has increased over the past two decades, climate change is still considered a low priority in the context of other issues American society is confronted with today (Bostrom et al., 1994; Read et al., 1994; Bord et al., 1998; Henry, 2000; Seacrast et al., 2000; Dessai et al., 2004; DEFRA, 2002, 2007; Leiserowitz, 2005, 2010; Ockwell et al., 2009). Americans regard both the environment and climate change as relatively low national priorities. For example, in a 2000 Gallup poll, the environment ranked 16th on Americans' list of most important problems facing the country today (Dunlap & Saad, 2001). Moreover, climate change ranked 12th out of 13 environmental issues, just below urban sprawl. Thus, Americans seem to be highly concerned about climate change as an individual issue, yet think it is less important than nearly all other national or environmental issues comparatively. Leiserowitz (2005) states that the low standing of climate change as a public concern reflects a widespread public perception that the issue is removed in space and time. Ockwell et al. (2009) adhere to the same conclusion, arguing that the American public believes climate change will primarily affect future generations and less developed countries. Furthermore, public concern for climate change is influenced by various uncertainties, public misconceptions, miscommunication, and competition for agenda seeking attention on an overwhelming socioenvironmental agenda (Kempton, 1991; Seacrest et al., 2000; Leiserowitz, 2005; Lorenzoni et al., 2005; Smith, 2005; Moser, 2006). As a result, the concept of "dangerous" climate change is not only contested among scientists and policy makers but among the American public as well.

Gaps between scientific and public understanding of climate change

The existing literature presents different explanations for the public dissensus over climate change, especially considering the broad consensus among the scientific community regarding the reality and risks of climate change. The predominant ways in which the public tends to think about the issue of climate change increases the likelihood of systematic misunderstandings (Weber & Stern, 2011). For example, people who rely on personal experience to determine the likelihood and level of threat of climate change can easily underestimate or overstate the real risks (Weber, 1997). Furthermore, due to the complexity and uncertainties of climate change, mental models are often incorrectly applied in the context of climate change (Bostrom et al., 1994). Instead of making judgments based on scientific evidence, decision-making processes are often driven by affect, values, or worldviews (Slovic, 1987). The different reasons for the controversy over climate change and the gap between scientific and public understanding, relevant to the research presented in this book, can be categorized into two different groups. These two groups are "lay mental models and misconceptions" and "worldviews and cultural values". The following paragraphs provide an in-depth discussion of the existing body of knowledge in these two areas.

Lay mental models and misconceptions

According to several studies, the public may not be totally aware of the causes of climate change and have misconceptions of what climate change is, many distrust the science of climate change, or believe it is not an urgent topic, but distant in time and space (Ungar, 2000; Hartley et al., 2011; Pidgeon & Fischoff, 2011). Surveys show that many Americans believe the climate change impacts other populations in other countries but not in the United States. Furthermore, only a small number among the US public connects climate change to direct health impacts. This demonstrates a clear gap between lay modes and expert assessments illustrated in the current IPCC reports (2013) or the report published by the US Global Change Research Program (Thomas et al., 2009), which focuses specifically on climate change impacts in the United States.

The existing body of knowledge, shows that lay mental models of climate change suffer from numerous misconceptions. This can be explained by misunderstandings of the science underlying climate change (Lorenzoni et al., 2005). According to Kempton (1991), new information on climate change is categorized by the public into four concepts or mental models. The most popular misbelief is that climate change is caused by increased ultraviolet light entering the atmosphere due to stratospheric ozone depletion. Although some interdependencies exist, these are only secondary, tertiary, or lesser effects. In addition, many public beliefs about ozone depletion are false or incomplete. For example, survey participants blamed aerosol for climate change despite the fact that they have been banned in the United States for decades. Whereas the ozone hole is a well-established concept in the American public consciousness, the greenhouse effect is only being recognized as a subset of the ozone hole phenomenon. Another popular misconception is that GHG emissions are just a form of air pollution (Brenchin, 2003; Lorenzoni et al., 2005). As a result, many people believe and support traditional pollution controls as the solution to increasing GHG emissions. However, actions such as filters and strengthening pollution controls alone do not stop climate change. The air pollution model focuses on industrial smokestacks and vehicle sources, which are a major source of GHG emissions.

Nonetheless, by applying this mental model to climate change, the public does not recognize the negative impact from seemingly nonpolluting sources such as farming, ranching, or leaking refrigeration (Kempton, 1991). The third concept, plant photosynthesis, also plays an important role in the public's misconception of climate change. The majority of survey participants in several studies (Bostrom et al., 1994; Read et al., 1994; Kempton, 1991, 1997; Henry, 2000) showed a sufficient understanding of the concept that trees absorb CO_2 and produce oxygen. This knowledge, combined with increasing media reports of forest destruction, led to the misconception that all atmospheric oxygen could be exhausted due to deforestation. Consequently, fighting deforestation is among the most popular policy responses to mitigate and adapt to climate change (Henry, 2000; Leiserowitz, 2006; DEFRA, 2007).

However, the contribution of growing plants to atmospheric oxygen is almost entirely offset by the decay of plants after their death. To increase atmospheric oxygen, dead plants would have to be buried before decomposing (Kempton, 1991). The fourth important misconception is that people underestimate the temperature change required for severe climate induced effects (Kempton, 1991, 1997; Seacrest et al., 2000; Dessai et al., 2004; Leiserowitz, 2006). To many Americans, an average temperature rise of less than 10°F does not seem very harmful, because they are familiar with high winter to summer temperature swings and major geographical differences in temperature. Because climate change impacts occur with small temperature changes, the public may not feel a high urgency to develop and support mitigation or adaptation strategies for climate change. Knowledge of the misconceptions and public imagery within a country is important in order to develop effective communication programs on an international and national scale.

Worldviews and cultural values

Worldviews and values can have a strong impact on how the risks and threats of climate change are perceived by the public (Slovic, 2000; Hulme, 2009; Kahan et al., 2011) and thus influence policy support and the willingness to commit to behavioral changes. Research in this field, however, is still very theoretical based and not many studies providing empirical data for validation are available at this point. The existing literature argues that perceptions of risks, such as those related to climate change, are socially constructed and can vary by culture, human development, affluence, national experience with risks, and demographics (Slovic, 2000). Furthermore, cultural theorists argue that our worldviews and our values play an important role in public risk perception and behavior (Douglas, 1966, 1970; Douglas & Wildavsky, 1982; Douglas et al., 1998).

More recently, several studies support the argument that an insufficient level of knowledge, the inability to assess technical information by the lay public, or the resulting reliance on inappropriate cognitive heuristics, do not explain the gap between the science and the public (Verwij et al., 2006; Kahan, 2010; Kahan et al., 2011; Weber & Stern, 2011). These studies acknowledge that public understanding of climate change needs improvement, but emphasize that the issue is not illiteracy among the lay public. Instead, people who doubt the reality of human–climate change and its negative impacts don't lack knowledge, but have a different understanding of the topic and thus interpret scientific results differently.

Worldwide, awareness of climate change is growing and is penetrating further into sociopolitical and cultural life. As a result, understanding how belief systems and perceptions impact public discussions of climate change and possible responses become increasingly important. Research suggests that disagreements about the issue of climate change exist because people view their responsibilities to future generations differently, value humans and nature in different ways, have different attitudes to climate risk, and are influenced by cultural cognition (Douglas & Wildavski, 1982). The theory of cultural cognition implies that the individual's risk perception of climate change is formed and reinforced by values that they have in common with others. Thus, proponents of this school of thought argue that the disagreements over climate change are in fact a conflict between groups that are separated by more general opposing perceptions of environmental and technological risks based on their members cultural outlooks (Verweij et al., 2006; Kahan et al., 2011).

A recent empirical study by Kahan et al. (2011) shows that high scientific literacy and numeracy among the lay public can enforce cultural polarization and widen the gap between social groups with opposing worldviews and values. The data suggest that people who dismiss the reality or dangers of climate change based on their values become even more dismissive. On the other hand, people who already believed in human-induced climate change and were concerned about possible negative impacts became even more concerned after being exposed to scientific literature. Overall, the study participants with high levels of scientific literacy were somewhat more likely to dismiss the seriousness of climate change compared to people with lower levels. Thus, instead of believing in climate change and supporting adaptation and mitigation policies, public misconceptions of climate change risks are most likely enforced as they become more knowledgeable. In order to improve the public's attitude towards climate change policies and their willingness to commit to behavioral changes, communication efforts cannot focus on presenting knowledge alone (NRC, 2005; Weber & Stern, 2011).

The results presented in Chapters 4 and 5 aim to advance the theory, through empirical evidence, that we disagree about climate change because we have different belief systems (Hulme, 2009) mediated through culture. Furthermore, the insights gained will improve the understanding to what degree mental models, scientific illiteracy and misconception, and cultural values effect climate change risk perception, behavior, and policy support.

Climate change communication

Although, the field of climate change communication research is still relatively young (NRC, 2010a), studies have already identified key aspects, guiding principles, and barriers to improving communication and education efforts (Leiserowitz, 2005; Smith, 2005; Frumkin & McMichael, 2008; Ockwell et al., 2009; Moser, 2010; Pidgeon & Fischhoff, 2011; Moloney & Strengers, 2014).

Guiding principles and barriers

One important shortcoming of past climate change communication efforts is that they are mainly based on the "information deficit model" (Irwin & Wynne, 1996), which assumes that people are "empty vessels" waiting to be filled with information that will propel them into rational action. However, these communication efforts do not take into account the heterogeneity of the public. Public groups can differ in their values and have diverse resources, which causes them to interpret and use information differently (Weber & Stern, 2011). Therefore, people disagree about climate change because communication programs fail to acknowledge that an individual's position to climate change represents certain values and worldviews that separate different cultural groups from one another (Kahan et al., 2011). Other research shows that global climate change communication programs fail, because they do not incorporate knowledge on risk perceptions, factors of trust, how science is conceived, moral issues, the role of uncertainty, the nature of the threat, and other factors related to the social and cultural dimensions of science (Wardekker, 2004).

This is not to say that education is not part of an effective public communication effort, but rather that it should be based on elements such as an understanding of individuals' existing knowledge, their concerns, worldviews and values and their abilities to react to the challenges of climate change. Moreover, disregarding these elements can increase the public dissensus over climate change and decrease support for climate change policies. Communication strategies should be designed with great caution since people tend to dismiss information, which is contrary to their worldview as a direct assault on their values and the competence of the persons they trust (Kahan et al., 2011).

In her study, "Talk of the city: engaging urbanities on climate change", Susanne Moser (2006) addresses questions about key audiences, appropriate messengers, framings and messages, reception of climate change information, and the choice of communication mediums and formats to achieve different communication and engagement goals. The author argues that past climate change communication efforts were not tailored towards a particular audience, but only focused on the science and overall impacts. Moser explains that editors, scientists, and policy makers alike always have to ask themselves who the audience is they are trying to communicate with. Moser shows that it is important to choose appropriate language and frames to talk about the issue of climate change and possible mitigation or adaptation policies and strategies. Moser defines effective climate change communication as "any form of public engagement that actually facilitates an intended behavioral, organizational, political and other social change consistent with identified mitigation and adaptation goals". Moser (2006) concludes that information or knowledge is not enough to change someone's behavior. Instead, the key challenge of effective communication is to motivate the public to begin and sustain the required behavioral changes.

Since many past communication strategies underperformed, more recent communication programs have started to emphasize the targeted population's guilt of causing climate change and their responsibility to take action against it (Gifford et al., 2011). Another promising communication approach is to incorporate the strategies and methods of social marketing (Corner et al., 2014). The central strategy of this technique is to divide the audience into different groups according to their values and attitudes to develop individual communication programs tailored to each group (Lazer & Kelley, 1973). This does not mean that education is not part of an effective public communication effort, but rather that it should be based on elements such as an understanding of individuals' existing knowledge, their concerns, worldviews and values, and their abilities to react to the challenges of climate change. Therefore, for improved or enhanced communication, broader and deeper knowledge of the public's risk perception, experiences, and attitudes towards climate change are needed.

The role of the mass media and the scientific community

The existing body of knowledge points to the mass media as a significant contributor to the current dissensus over climate change, especially in the United States (Nelkin 1995; Boykoff, 2008; Leiserowitz, 2005; Smith, 2005; Antilla, 2005, 2010). Mass media, however, is simultaneously also considered a key part of successfully communicating climate change and increase public policy support. Research shows that the way the public perceives climate change is strongly influenced by how and to what degree the media communicates the existing scientific knowledge (Mazur & Lee, 1993; Wilkins, 1993; Mormont & Dasnoy, 1995; Trumbo, 1996; Brulle et al., 2010).

Leiserowitz (2005) links the recent decline of public concern over climate change to the way climate change is presented in the mass media. His and similar studies point out that since 1988, when climate change was a front-page story, television network coverage declined by 50 percent and national newspaper coverage dropped by 25 percent (Frame Works Institute, 2001). The severe drought, the heat wave, and James Hansen's testimony before Congress in 1988 led to a dramatization and amplification of the topic by the media and environmental groups, and concern peaked in 1998 (Bord et al., 1998). Simultaneously, scientific journals increasingly emphasize the uncertainty in climate change predictions and public interest faded with the onset of cooler, wetter summers (Ungar, 1992; Smith, 2005).

Not only the amount of media coverage is important, however, but the way climate change is presented is significant as well. The different sources for potential shortcomings of climate change communication by the media can be traced back to the actual professional norms journalists rely on (Boykoff & Boykoff, 2004). These norms can be grouped into first- and second-order journalistic norms and can lead to the misrepresentation of the science behind climate change and thus spark an informational bias regarding anthropogenic climate change. First-order journalistic norms include personalization, dramatization, and novelty, whereas secondary norms consist of authority-order and balance.

According to Gans (1979), due to the first-order journalistic norm of personalization the media tends to focus on individuals instead on group dynamics or social processes. Therefore, in the context of climate change, the media focus on individuals affected by negative, shifts the public attention only to a small part of climate change enforcing the general public's believe that climate change is an issue removed in time and space. Furthermore, by focusing on short-term events and often disregarding the causes or long-term trends, the media encourages public misconceptions and skepticism towards anthropogenic climate change (Boykoff & Boykoff, 2004).

The second first-order norm, dramatization, is also very important for understanding the media's reporting on climate change. In order to increase reader-or, the media tends to focus on only current and highly visible crisis instead of providing a broader and complete representation of the issue, the causes, or the solutions (Wilkins & Paterson, 1987; Sheppard, 2012). As previously discussed in Chapter 4, the survey data show that people tend to believe that climate change is in general a serious threat but not necessarily to themselves, but to plants and animals as well as to people in other countries. The media often disregards climate change impacts that are less visible or dramatic because of the lack of excitement or controversy. Simultaneously, the limited media coverage addressing potential solutions is often rather simplistic focusing on high profile policies such as wind turbines or electric vehicles, while disregarding lesser known solutions with similar or even higher benefits. In addition the often complicated scientific language behind climate change and high degrees on uncertainties makes it very difficult for journalists to report on climate change while conforming to the dramatization norm (Ungar, 2000). As a result, the incomplete coverage of climate change makes it difficult for the public to recognize the connections between the impacts and causes, as well as positive solutions they should be considering. Instead, climate change is often perceived as an issue removed in time and space, which does not require immediate action.

The final first-order norm, novelty, also represents significant barrier to adequate and comprehensive climate change communication by the media (Wilson, 2000). Journalists are always looking for the new and breaking story, which results in a preference for covering crisis instead of chronic social or environmental problems such as climate change. Thus, the actual causes or long-term consequences are often disregarded in today's 24-hour news cycle. Furthermore, since climate change is a slowly evolving trend of which many of the impacts are not visible yet, it seldom is considered prime-time news material (Boykoff, 2011). All three of these first-order norms enforce the second-order journalistic norms, authority-order, and balance, which also pose significant barriers to unbiased reporting on climate change. When dealing with complex issues, journalists often rely on the opinions of high-profile figures such as government officials, business leaders, scientists, and others. This can lead to a so-called authority-order bias where journalist may be relying on experts with their own agendas or even conflicting point of views. In the context of climate change, this can result in the unjustified diffusion or amplification of public concern influencing public trust in authority figures and policy decision making (Pidgeon & Gregory, 2004; Lorenzoni & Pidgeon, 2006). Therefore, journalists need to be cautious when dealing with their sources that they don't become agenda-builders for different interest groups who are trying to use the media as a delivery vehicle for their own communication objectives.

Probably the most influential norm that determines the type of climate change coverage by the media is balanced reporting. Unfortunately, this second-order journalistic norm has a significant impact on the public's perception of climate change and enforces the public misconception that the reality and dangers of climate change is still highly debated within the scientific community. In general, the goal of balanced reporting is to ensure unbiased reporting by giving equal attention to the arguments of all conflicting parties involved (Entman, 1989). However, in the context of climate change, this norm can be a substantial barrier to successful and objective communication. By focusing on a balanced coverage of climate change the media gives even consideration to the arguments of skeptics as to the overwhelming scientific body of evidence that supports anthropogenic climate change (Nelkin, 1995). For example, a study conducted in the United States (Boykoff & Boykoff, 2004) shows that over a 15-year period the majority of the media reports about climate change gave roughly equal attention to the two opposing arguments that (a) climate change is caused by human behavior and (b) that natural functions alone can explain the rise in the average temperature. As a result, the media enforces the public misconception that the reality and dangers of climate change is still highly debated within the scientific community (Boykoff, 2008). This presents a significant barrier to public's willingness to fully commit to behavioral changes or support climate change policies (Antilla, 2010) and thus needs to be addressed in future communication efforts.

Furthermore, Smith (2005) points out that the notions of danger caused by climate change are significantly mediated by news and other broadcast and published sources. Smith argues that the scientific community and policy makers need to be more aware and critical of how climate change is portrayed in the media. The author criticizes that scientists and policy specialists are seemingly concerned and reluctant to present their arguments in a news context. Smith believes that scientists are afraid of losing credibility through simplification by the news stations, giving up control of statement to editors, and the fear that their amount of work is not being recognized in a short 2-minute news segments. Furthermore, Smith points out an important shortcoming for presenting climate change in the way news stories are ordered during a broadcast. The author states that the organization of topics from local to national to scales and by subject categories makes it difficult for editors to place climate change. As climate change is characterized by impacting and interacting on and between all spatial scales and various categories, the topic is usually placed by editors at a global scale. As a result of declining, and inappropriately balanced and conceptualized media coverage, the public is inclined to underestimate possible negative impacts on the local scale and therefore do not support adaptation and mitigation strategies to the degree necessary.

Furthermore, the scientific community has failed as well to communicate the effects of climate change in a comprehensive and easy-to-understand manner to the public or the media (Sheppard, 2012). Scientists publish their work in scientific journals which are full of terminology the public is unfamiliar with and heavy with information that is often abstract, complex, remote, depressing, and at times overwhelming. Moreover the way scientists communicate their findings, for example through journal articles, conferences, or reports do usually not allow interaction or querying by the lay public. For the public, this makes personal connections and engagement difficult. This study aims, through thematic literature review and survey analysis, to identify perceptional factors which need to be considered in future communication efforts to reduce current misconceptions, change attitudes, and increase support for climate change policies.

Bibliography

Akerlof K., Maibach E.W., Fitzgerald D., Cedeno A.Y., Neuman A. (2013). Do people "personally experience" global warming, and if so how, and does it matter? *Global Environmental Change 23*, 81–91.

Antilla, L. (2005). Climate of scepticism: US newspaper coverage of the science of climate change. *Global Environmental Change, 15*(4), 338–352.

Antilla, L. (2010). Self-censorship and science: A geographical review of media coverage of climate tipping points. *Public Understanding of Science, 19*(2), 240–256.

Babbie, E. (2007). *The Practice of Social Research*. Belmont, CA: Thomson Wadsworth.

Baggaley, A., & Hull, A. (1983). The effect of nonlinear transformations on a Likert scale. *Evaluation & the Health Professions, 6*, 483–491.

Barkham, R. (2014). Investing in resilience: Ranking the most resilient cities. *Urbanland Magazine*. http://urbanland.uli.org/sustainability/investing-resilient-cities/ (last accessed 1/23/2015)

Booth, W. (2010). Mexico seeks leading role in climate policy. *The Washington Post*. http://www.washingtonpost.com/wpdyn/content/article/2010/11/28/AR2010112802975.html (last accessed 10/24/2014)

Bord, R.J., Fisher, A., & O'Connor, R.E. (1998). Public perceptions of global warming: United States and international perspectives. *Climate Research, 11*, 75–84.

Bostrom, A., Granger Morgan, M., Fischhoff, B., & Read, D. (1994). What do people know about global climate change? *Risk Analysis, 14*(6), 959–970.

Bowonder, B., Kasperson, J.X., & Kasperson, R.E. (1985). Avoiding future bhopals. *Environment: Science and Policy for Sustainable Development, 27*(7), 6–37.

Boykoff, M. (2008). Media and scientific communication: A case of climate change. Geological Society, *Special Publication, 305*, 11–18.

Boykoff, M. (2011). *Who Speaks for the Climate? Making Sense of Media Reporting on Climate Change*. New York, NY: Cambridge University Press.

Boykoff, M., & Boykoff, J. (2004). Balance as bias: Global warming and the US prestige press. *Global Environmental Change, 14*(2), 125–136.

Brechin, S.R. (2003). Comparative public opinion and knowledge on global climate change and the Kyoto Protocol: The U.S. versus the world? *International Journal of Sociology and Social Policy, 23*(10), 106–134.

Brown, J.D. (2011). Likert items and scales of measurement? *SHIKEN: JALT Testing & Evaluation SIG Newsletter, 15*(1), 10–14.

Brulle, R.J., Carmichael, J., & Jenkins J.C. (2012). Shifting public opinion on climate change: An empirical assessment of factors influencing concern over climate change in the U.S., 2002–2010. *Climatic Change, 114*, 169–188.

Burton, I., Kates, R.W., & White, G.F. (1978). *The Environment as Hazard*. New York, NY: Oxford University Press.

Corner, A., Markowitz, E., & Pidgeon, N. (2014). Public engagement with climate change: The role of human values. *WIREs Climate Change, 5*(3), 411–422.

Curry, J. (2011a). Reasoning about climate uncertainty. *Climatic Change, 108*, 723–732.

Curry, J. (2011b). Anticipating the Climate Black Swan. Retrieved from http://judithcurry.com/2011/05/02/anticipating-the-climate-black-swan/ (last accessed 3/20/14)

Department for Environment, Food and Rural Affairs (DEFRA). (2002). *Survey of public attitudes to quality of life and to the environment: 2001*. London, UK: Author.

DEFRA. (2007). *Survey of Public Attitudes to Quality of Life and to the Environment: 2006*. London, UK: Author.

Department of Health (UK). (1997). *Communicating about Risks to the Public Health: Pointers to Good Practice*. London, UK: Author.

Dessai, S., Adger, W., Hulme, M., Turnpenny, J., Koehler, J., & Warren, R. (2004). Defining and experiencing dangerous climate change. *Climatic Change, 64*(1), 11–25.

Dillman, D.A., Smyth, J.D., & Christian, L.M. (2009). *Internet, Mail, and Mixed-Mode SURVEYS: The Tailored Design Method*. Hoboken, NJ: Wiley.

Douglas, M. (1966). *Purity and Danger: An Analysis of Concepts of Pollution and Taboo*. London, UK: Taylor.

Douglas, M. (1970). *Natural Symbols: Explorations in Cosmology*. London, UK: Barrie and Rockliff.

Douglas, M., Gasper, D., Ney, S., & Thompson M. (1998). Human needs and wants. In S. Rayner and E.L. Malone (Eds.), *Human Choice and Climate Change* (pp. 195–265). Columbus, OH: Battelle Press.

Douglas, M, & Wildavsky, A. (1982). *Risk and Culture: An Essay on the Selection of Technological and Environmental Dangers*. Berkeley: University of California Press

Dunlap, R., & Saad, L. (2001). *Only One in Four Americans Are Anxious about the Environment*. Washington, DC: Gallup World Headquarters.

Edwards, W. (1961). Behavioral decision theory. In P.R. Farnsworth, O. McNemar, & Q. McNemar (Eds.), *Annual Review of Psychology* (pp. 473–498). Palo Alto, CA: Annual Reviews, Inc.

Entman, R.W. (1989). *Democracy without Citizens: Media and the Decay of American Politics*. New York, NY: Oxford University Press.

Field, A. (2013). *Discovering Statistics Using SPSS*. London: Sage.

Finucane, M.L., Alhakami, A., Slovic, P., & Johnson, M. (2000). The affect heuristic in judgments of risks and benefits. *Journal of Behavioral Decision Making, 13*(1), 1–17.

Fischer, A., Peters, V., Neebe, M., Vavra, J., Kriel, A., Lapka, M., Megysi, B. (2012). Climate change? No, wise resource use is the issue: Social representations of energy, climate change and the future. *Environmental Policy and Governance, 22*, 161–176.

Fischhoff, B., Slovic, P., Lichtenstein, S., Read, S., & Combs, B. (1978). How safe is safe enough? A psychometric study of attitudes towards technological risks and benefits. *Policy Sciences, 9*, 127–152.

Flynn, J., Chalmers, J., Easterling, D., Kasperson, R.E., Kunreuther, H., Mertz, C.K., . . . Slovic, P. (1995). *One Hundred Centuries of Solitude: Redirecting America's High-Level Nuclear Waste Policies*. Boulder, CO: Westview Press.

Fowler Jr., F.J. (2008). *Survey Research Methods: Applied Social Research Methods Series, No. 1*. Thousand Oaks, CA: Sage..

Frame Works Institute. (2001). *Talking Global Warming*. Washington, DC: Frame Works Institute.

Freudenberg, W.R., & Rursch, J.A. (1994). The risks of "putting the numbers in context": A cautionary tale. *Risk Analysis, 14*(6), 949–958.

Frumkin, H., & McMichael. (2008). Climate change and public health: Thinking, communicating, acting. *American Journal of Preventive Medicine, 35*(5), 403–410.

Gaerling, T., & Golledge, R.G. (1993). *Behavior and Environment: Psychological and Geographical Approaches*. Amsterdam, The Netherlands: Elsevier Science Publishers.

Gans, H. (1979). *Deciding What's News*. Pantheon, NY: Northwestern University Press

Gifford, R., Kormos, C., & McIntyre, A. (2011). Behavioral dimensions of climate change: Drivers, responses, barriers, and interventions. *WIREs Climate Change, 2*(6), 801–827.

Gomez, L.S., Jenkins-Smith, H.C., & Miller, K.W. (1992). *Changes in Risk Perception over Time*. Washington, DC: Department of Energy.

Hartley, L.M., Wilke, B.J., Schramm, J.W., D'Avanzo, C., & Anderson, C.W. (2011). College students' understanding of the carbon cycle: Contrasting principle-based and informal reasoning. *BioScience, 61*(1), 65–75.

Henry, A.D. (2000). Public perception of global warming. *Human Ecology Review, 7*(1), 25–30.

Hill, S. (2011). Economic powerhouse Germany: High-tech niche ascendency and economic democracy propel the export star. *IP Global Edition, 2*, 7–12.

Hinchliffe, S. (1996). Helping the earth begins at home: The social construction of socio-environmental responsibilities. *Global Environmental Change, 6*(1), 53–62.

Hulme, M. (2009). *Why We Disagree about Climate Change: Understanding Controversy Inaction and Opportunity*. New York: Cambridge University Press.

IPCC (2007). *Climate Change 2007: Fourth Assessment Report of the Intergovernmental Panel on Climate Change*. Cambridge, UK: Intergovernmental Panel on Climate Change.

IPCC. (2013). *Climate Change 2013: The Physical Science Basis. Contribution of Working Group I to the fifth Assessment Report of the Intergovernmental Panel on Climate Change.* Cambridge, UK/New York, NY: Cambridge University Press.

Irwin, A., & Wynne, B. (Eds.). (1996). *Misunderstanding Science? The Public Reconstruction of Science and Technology.* Cambridge, UK: Cambridge University Press.

Jaccard, J., & Wan, C.K. (1996). *LISREL Approaches to Interaction Effects in Multiple Regression.* Thousand Oaks, CA: Sage.

Jamison, S. (2004). Likert scales: How to (ab)use them. *Medical Education, 38,* 1212–1218.

Jin, M.H., & Shriar, A.J. (2013). Linking environmental citizenship and civic engagement to public trust and environmental sacrifice in the Asian context. *Environmental Policy and Governance, 23,* 259–273.

Johnson, B.B., & Slovic. P. (1995). Presenting uncertainty in health risk assessment: Initial studies of its effects on risk perception and trust. *Risk Analysis, 15*(4), 485–494.

Kahan, D.M. (2012). Cultural Cognition as a conception of the cultural theory of risk. In S. Roesser, R. Hillerbrand, & M. Pesterson (Eds.). *Handbook of Risk Theory: Epistemology, Decision Theory, Ethic and Social Implications of Risk* (pp. 726–759). London: Springer.

Kahan, D.M., Peters, E., Braman, D., Slovic, P., Wittlin, M., Ouellette, L.L., & Mandel, G. (2011). The tragedy of the risk-perception commons: culture conflict, rationality conflict, and climate change Cultural Cognition Project. *Cultural Cognition Project, Working Paper No. 89.*

Karl, T. R., Melillo, J. M., & Peterson, T. C. (Eds.). (2009). *Global Climate Change Impacts in the United States.* New York, NY: Cambridge University Press.

Kasperson, J.X., Kasperson, R.E., & Turner II, B.L. (Eds.). (1995). *Regions at Risk: Comparisons of Threatened Environments.* New York, NY: United Nations University Press.

Kasperson, R.E., Renn, O., Slovic, P., Brown, H.S., Emel, J., Goble, R., . . . Ratick, S. (1988). The social amplification of risk: A conceptual framework. *Risk Analysis, 8*(2), 177–187.

Kasperson, R., Golding, D., & Tuler, S. (1992). Social distrust as a factor in sitting hazardous facilities and communicating risks. *Journal of Social Issues, 48*(4), 161–187.

Kates, R.W., Hohenemser, C., & Kasperson, J.X. (1985). *Perilous Progress: Managing the Hazards of Technology.* University of Minnesota: Westview Press.

Kempton, W. (1991). Lay perspectives on global climate change. *Global Environmental Change, 1*(3), 183–208.

Kempton, W. (1997). How the public views climate change. *Environment and Behavior, 39*(9), 12–21.

Krimsky, S., & Golding, S. (1992). *Social Theories of Risk.* Westport, CT: Praeger Publishers.

Labovitz, R.W. (1975). Comment on the Henkel's paper: The interplay between measurement and statistics. *Pacific Sociological Review, 18,* 27–35.

Lazer W., & Kelly, E.J. (1973). *Social Marketing Perspectives and Viewpoints.* Dorsey: Ontario.

Leiserowitz, A. (2005). American risk perceptions: Is climate change dangerous? *Risk Analysis, 25*(6), 1433–1442.

Leiserowitz, A. (2006). Climate change risk perception and policy preferences: The role of affect, imagery, and values. *Climatic Change, 77*(1), 45–72.

Leiserowitz, A. (2010). Risk perception and behavior. In S.H. Schneider, A. Rosencranz, M.D. Mastrandrea, & Kuntz-Duriseti, K. (Eds.), *Climate Change Science and Policy* (pp. 175–184). Washington, DC: Island Press.

Leiserowitz, A., Maibach, E., & Roser-Renouf, C. (2010). *Climate Change in the American Mind: Americans' Global Warming Beliefs and Attitudes in January 2010.* New Haven, CT: Yale University and Mason University.

Loewenstein, G., & Mather, J. (1990). Dynamic processes in risk perception. *Journal of Risk and Uncertainty, 3,* 155–175.

Loewenstein, G., Weber, E.U., Hsee, C.K., & Welch, N. (2001). Risk as feelings. *Psychological Bulletin, 127*(2), 267–286.

Lorenzoni, I., & Pidgeon N. (2006). Public views on climate change: European and USA perspectives. *Climate Change, 77*, 73–95.

Lorenzoni, I., Pidgeon, N., & O'Connor, R. (2005). Dangerous climate change: The role for risk research. *Risk Analysis, 25*(6), 1387–1398.

Maibach, E., Roser-Renouf, C., & Leiserowitz, A. (2009). *Global Warming's Six Americas 2009: An Audience Segmentation Analysis.* http//www.climatechangecommunication.org/images/files/GlobalWarmingsSixAmericas2009c.pdf (last accessed 1/23/2015)

Malhi, Y., Roberts, J.T., Betts, R.A., Killeen, T.J., Li, W., & Nobre. C.A. (2008). Climate change, deforestation, and the fate of the Amazon. *Science, 319*(5860), 169–172.

Maurer, J., & Pierce, H.R. (1998). A comparison of Likert scale and traditional measures of self-efficacy. *Journal of Applied Psychology, 83*, 324–329.

Mazur, A,. and Lee, J. (1993). Sounding the global alarm: Environmental issues in the US national news. *Social Studies of Science, 23*(4), 681–720.

Moloney S., & Strengers Y. (2014). "Going green?": The limitations of behaviour change programmes as a policy response to escalating resource consumptions. *Environmental Policy and Governance 24*, 94–107.

Mormont, M., & Dasnoy, C. (1995). Source strategies and the mediatization of climate change. *Media, Culture and Society, 17*, 49–64.

Moser, S.C. (2006). Talk of the city: Engaging urbanities on climate change. *Environmental Research Letters, 1*, 1–10.

Moser, S.C. (2010). Communicating climate change: History, challenges, process and future directions. Wiley Interdisciplinary Reviews. *Climate Change, 1*(1), 31–53.

National Research Council. (2005). *How Students Learn: History, Mathematics, and Science in the Classroom.* Washington, DC: The National Academies Press.

National Research Council (NRC). (2010a). *Informing an Effective Response to Climate Change.* Washington DC: The National Academies Press.

NRC. (2010b). *Adapting to the Impacts of Climate Change.* Washington, DC: The National Academies Press.

Nelkin, D. (1995). *Selling Science: How the Press Covers Science and Technology, revised ed.* New York, NY: W.H. Freeman & Co.

Nisbett, R., & Ross, L. (1980). *Human Inference: Strategies and Shortcomings of Social Judgment.* Englewood Cliffs, NJ: Prentice-Hall.

Nye Jr., J.S., Zelikow, P.D., & King, D.C. (1997). *Why People Don't Trust Government.* Cambridge, UK: Harvard University Press.

Ockwell, D., Whitmarsh, L., & O'Neill, S. (2009). Reorienting climate change communication for effective mitigation: Forcing people to be green or fostering grass-roots engagement? *Science Communication, 30*(3), 305–327.

O'Connor, R., Bard, R., & Fisher, A. (1999). Risk perceptions, general environmental beliefs, and willingness to address climate change. *Risk Analysis, 19*(3), 461–471.

O'Leary, Z. (2004). *The Essential Guide to Doing Research.* Thousand Oaks, CA: Sage.

O'Neill, S., & Hulme, M. (2009). An iconic approach for representing climate change. *Global Environmental Change, 19*, 402–410.

Owuor, C.O. (2001). *Implications of Using Likert Data in Multiple Regression Analysis.* Vancouver, Canada: University of British Columbia.

Paton, D. (2008). Risk perception and volcanic hazard mitigation: Individual and social perspectives. *Journal of Volcanology and Geothermal Research, 172*, 179–188.

Peters, R.G., Covello, V.T., & McCallum, D.B. (1997). The determinants of trust and credibility in environmental risk communication. *Risk Analysis, 17*(1), 43–54.

Pew Research Center. (2010). *Who's Winning the Clean Energy Race? Growth, Competition and Opportunity in the World' Largest Economies.* Washington, DC: The Pew Charitable Trusts.

Pidgeon, N., & Fischhoff, B. (2011). The role of social and decision sciences in communicating uncertain climate risks. *Nature Climate Change, 1,* 35–41.

Pidgeon, N.F., & Gregory, R. (2004). Judgment, decision-making and public policy. In Koehler, D., & Harvey, N. (Eds.), *Handbook of Judgment and Decision-Making* (pp. 604–623). Oxford, UK: Blackwell.

Pijawka, K.D., & Mushkatel, A.H. (1991). Symposium on the development of nuclear waste policy: siting the high-level nuclear waste repository. *Review of Policy Research, 10*(4), 88–89.

Read, D., Bostrom, A., Granger Morgan, M., Fischhoff, B., & Smuts, T. (1994). What do people know about global climate change? 2. Survey Studies of educated laypeople. *Risk Analysis, 14*(6), 971–982.

Renn, O., & Levine, D. (1991). Credibility and trust in risk communication. In R.E. Kasperson and P.J.M. Stallen (Eds.), *Communicating Risks to the Public* (pp. 175–218). Dordrecht, The Netherlands: Kluwer Academic.

Rosenstone, S., Kinde, D., & Miller, W. (1997). *American National Election Study.* Ann Arbor, MI: Interuniversity Consortium for Political and Social Research.

Rotter, J. (1966). Generalized expectancies for internal versus external control of reinforcements. *Psychological Monographs, 80,* 1–28.

Seacrest, S., Kuzelka, R., & Rick, L. (2000). Global climate change and public perception: The challenge of translation. *Journal of the American Water Resources Association, 36*(2), 253–263.

Sheppard, S.R.J. (2012). *Visualizing Climate Change: A guide to Visual Communication of Climate Change and Developing Local Solutions.* London, UK: Routledge.

Short, J.F. (1984). The social fabric at risk: Toward the social transformation of risk analysis. *American Sociological Review, 49,* 711–725.

Simmons, W.M. (2007). *Participation and Power: Civil Discourse in Environmental Policy Decisions.* Albany: State University of New York Press.

Sjoeberg, L. (1998). World views, political attitudes and risk perception. *Risk-Health, Safety and Environment, 9,* 137–152

Slovic, P. (1986). Informing and educating the public about risk. *Risk Analysis, 6*(4), 403–415.

Slovic, P. (1987). Perception of risk. *Science, 236*(4799), 280–285.

Slovic, P. (1997). Trust, emotions, sex, politics, and sciences: Surveying the risk assessment battlefield. In M.H. Bazerman, D.M. Messick, A.E. Tenbrunsel, & K.A. Wade-Benzoni (Eds.), *Environment, Ethics, and Behavior* (pp. 277–313). San Francisco, CA: New Lexington.

Slovic, P. (2000). *The Perception of Risk.* London, UK: Earthscan.

Slovic. P (2010). *The Feeling of Risk.* London, UK: Earthscan.

Slovic, P., Fischhoff, B., & Lichtestein, S. (1981). Perceived risk: Psychological factors and social implications. In F. Warner & D.H. Slater (Eds.), *The Assessment and Perception of Risk* (pp. 17–34). London, UK: The Royal Society.

Slovic, P., Fischhoff, B., & Lichtenstein, S. (1984). Behavioral decision theory perspectives on risk and safety. *Acta Psychologica, 56,* 183–203.

Smith, J. (2005). Dangerous news: Media decision making about climate change risk. *Risk Analysis, 25*(6), 1471–1482.

Taleb, N.M. (2010). *The Black Swan: The Impact of the Highly Improbable.* Random House Trade Paperbacks.

Tate, R.B., Fernandez, N., Yassi, A., Canizares, M., Spiegel, J., & Bonet, M. (2003). Change in health risk perception following community intervention in Central Havana, Cuba. *Health Promotion International, 18*(4), 279–286.

Teixera, M. (2012). Mexico's climate law to face challenge under new president. *Reuters* http://www.reuters.com/article/2012/07/24/us-mexico-climate-policy-idUSBRE 86N0A22012072 4 (last accessed 10/24/2014)

Trumbo, C. (1996). Constructing climate change: Claims and frames in US news coverage of an environmental issue. *Public Understanding of Science, 5*, 269–283.

Tversky, A., & Kahneman, D. (1981). The framing of decisions and the psychology of choice. *Science, 211*, 453–458.

Ungar, S. (1992). The risen and (relative) decline of global warming as a social problem. *The Sociological Quarterly, 33*(4), 483–501.

Ungar, S. (2000). Knowledge, ignorance and the popular culture: Climate change versus the ozone hole. *Public Understanding of Science, 9*(3), 297–312.

Verweij, M., Douglas, M., Ellis, R., Engle, C., Hendriks, F., Lohmann, S., . . . Thompson, M. (2006). Clumsy solutions for a complex world: The case of climate change. *Public Administration, 84*, 817–843.

Vickers, A. (1999). Comparison of an ordinal and a continuous outcome measure of muscle soreness. *International Journal of Technology Assessment in Health Care, 15*, 709–716.

Wardekker, J.A. (2004). *Risk Communication on Climate Change*, PhD dissertation, Utrecht, The Netherlands: Utrecht University.

Weber, E.U. (1997). Perception and expectation of climate change. In M. Bazerman, D. Messick, A. Tenbrunsel, & K. Wade-Benzoni (Eds.), Psychological perspectives to environmental and ethical issues in management (pp. 314–341). San Francisco, CA: Jossey-Bass.

Weber, E.U., & Stern, P.C. (2011). Public understanding of climate change in the United States. *American Psychologist, 66*(4), 315–328.

Weidner, H., & Mez, L. (2008). German climate change policy: A success story with some flaws. *The Journal of Environment and Development, 17*(4), 356–378.

Whitmarsh, L. (2009). Behavioral responses to climate change: Asymmetry of intentions and impacts. *Journal of Environmental Psychology, 29*, 13–23.

Whitmarsh, L., O'Neill, S., Seyfang, G., & Lorenzoni, I. (2009). Carbon capability: What does it mean, how relevant is it, and how can we promote it? *Tyndall Working Paper, No. 132.*

Whitmarsh L, Seyfang G, & O'Neill S. (2011). Public engagement with carbon and climate change: to what extent is the public 'carbon capable'? *Global Environmental Change, 21*(1), 56–65. DOI: 10.1016/j.gloenvcha.2010.07.011

Wilkins, L. (1993). Between facts and values: Print media coverage of the greenhouse effect, 1987–1990. *Public Understanding of Science 2*, 71–84.

Wilkins, L., & Patterson, P. (1987). Risk analysis and the construction of news. *Journal of Communication, 37*(3), 80–92.

Wilson, K.M. (2000). Communicating climate change through the media: Predictions, politics, and perceptions of risks. In S. Allan, B. Adam, C. Carter (Eds.), *Environmental Risks and the Media* (pp. 201–217). New York, NY: Routledge.

Winston, A. (2010). The competing black swans of sustainability. *Harvard Business Review.* https://hbr.org/2010/09/the-competing-black-swans-of-s.htm 4 (last accessed 10/24/2014)

World Bank. (2014a). CO2 emissions (kt). http://data.worldbank.org/indicator/EN.ATM. CO2E.KT/countries?order=wbapi_data_value_2010+wbapi_data_value+wbapi_data_ value-last&sort=desc (last accessed 10/24/2014)

World Bank. (2014b). *World Development Indicators*. http://databank.worldbank.org/data/ views/reports/tableview.aspx?isshared=true (last accessed 10/24/2014)

Yin, R.K. (1994). *Case Study Research: Design and Methods*. London, UK: Sage.

4　Public perceptions of climate change

This chapter presents the first stage of the survey data analysis while addressing the first underlying research question of the "Global Survey on Public Attitudes towards Climate Change" research project:

- What are the public perceptions of climate change in terms of threat and risk, saliency of the issue, trust in climate change information, and acceptable public strategies?

By examining basic frequencies and significant country differences, the following paragraphs address how the public perceives the issue of climate change at the international level by looking at survey results from the nine countries: Canada, the United States, Mexico, Brazil, Spain, the Netherlands, Germany, the United Kingdom as a whole, and Japan. In particular, using descriptive statistical analysis methods, this section takes a close look at how the survey participants responded to questions dealing with the potential threats and risks of climate change, the saliency of the issue, trust in climate information and their sources, and possible policies and strategies which require the public's support.

By identifying key differences and similarities among countries and relationships between important variables, the descriptive exploration of the survey data provide the foundation for more sophisticated statistical analytical methods necessary for answering many of the remaining research questions. In addition, since policy makers do not have a good understanding of where various publics stand on climate change strategies, policies, priorities and what will be acceptable and supportive, the analyzed survey data in this chapter can assist policy makers in evaluating the appropriate choices to make and what would be seen as publically acceptable decision making.

Political saliency and public climate change concern

An important aspect of the public perceptions of global climate change is where people position that issue in the context of other problem areas the government focuses on. The idea of the relative importance of where climate change is ranked among all the other socioeconomic problem areas confronting populations is known as *political saliency* – how important the problem is for government to act on. One question in the survey instrument measured political saliency by asking the participants to indicate how important it is for government to act on nine separate problem areas. One of these nine areas was "reducing global climate change". The participants were asked to rank the nine issues by rating the level of importance for the government to act on a 4-point Likert scale. The scale was coded to analyze the answers as categorical data.

Past studies show undertaken in the United States show that the American public regards both the environment and climate change as relatively low national priorities (Leiserowitz, 2005; Ockwell et al., 2009; Leiserowitz et al., 2010). Project data support the existing body of knowledge on the global scale that climate change is considered a low saliency issue. Table 4.1 illustrates the scale, frequency distribution, and means for all nine countries combined.

Table 4.1 Global trends for political saliency of climate change

	Unimportant	Low Importance	Important	Very Important	Mean
Lowering the rate of violent crime	1.3%	4.1%	32.9%	63.7%	**3.56**
Improving the nation's schools	1.1%	5.2%	34.0%	59.8%	**3.52**
Reducing poverty	1.2%	6.1%	36.5%	56.2%	**3.48**
Increasing employment	1.1%	2.9%	29.7%	66.4%	**3.61**
Reducing global climate change	5.4%	15.1%	41.2%	38.3%	**3.12**
Improving air and water quality	1.8%	12.2%	45.1%	40.9%	**3.25**
Preventing global terrorism	2.6%	12.8%	39.8%	44.8%	**3.27**
Eliminating illegal drugs	3.8%	15.2%	37.0%	44.0%	**3.21**
Developing a comprehensive clean energy policy	2.9%	13.0%	42.5%	41.5%	**3.23**

Note: Survey participants were asked on a scale from 1 to 4, where 1 is "unimportant" and 4 is "very important", what level of importance should the government place on each of the nine problem areas.

With a mean of 3.12 (std. deviation ±.857), the results show that climate change is the *least* salient issue for the combined sample. Only 38.3 percent found it a "very important" issue for government to act on, rating it far below violent crime, schools, employment, and poverty. The combined results of the "level of concern" and the "political saliency" factors suggests that people believe that climate change is important, but certainly not a pressing issue. However, when you combine the two categories "important" and "very important", almost 80 percent of the participants want government to be involved in climate change policies to prevent and lessen its impacts.

Looking at the specific country data individually, reducing climate change ranks in the bottom third of identified problems despite the fact that the majority supported some form of governmental action. In the United Kingdom, the United States, and the Netherlands, it ranked last with the highest being the United States at 15.7 percent. When combining the "unimportant" and "low importance" categories, these three countries stand out as well. In the Netherlands, 36.7 percent of survey participants believe that the government should not address climate change at all or only as a low priority issue, followed by the United States with 34.4 percent and the United Kingdom with 25.8 percent. Based on the same 4-point Likert scale as the previous table, Table 4.2 shows the percentages, the mean values, and standard deviations, for each of the nine countries individually regarding the level of importance the government should place on reducing climate change. The table emphasizes the countries that stand out on either ends of the scale.

Except for the Brazilian sample, less than 50 percent of the participants in all other countries indicated that the government should treat climate

Table 4.2 Political saliency of climate change by country

	Unimportant	Low Importance	Important	Very Important	Mean	Std. Dev.
Brazil	2.4%	7.3%	31.4%	59.0%	**3.47**	**.73**
Mexico	3.9%	7.7%	38.6%	49.8%	**3.34**	**.78**
Spain	3.3%	10.5%	43.4%	42.9%	**3.26**	**.77**
Germany	3.5%	11.9%	41.7%	42.8%	**3.24**	**.80**
Japan	2.3%	7.7%	53.1%	32.9%	**3.17**	**.72**
Canada	3.7%	16.0%	42.3%	38.0%	**3.15**	**.82**
United Kingdom	6.3%	19.5%	43.5%	30.7%	**2.99**	**.87**
United States	15.7%	18.9%	35.4%	30.0%	**2.80**	**1.04**
Netherlands	5.2%	31.5%	42.0%	21.2%	**2.79**	**.83**

Note: Survey participants were asked on a scale from 1 to 4, where 1 is "unimportant" and 4 is "very important", what level of importance should the government place on reducing climate change?

change as a very important policy priority. This is followed by Mexico with 49.8 percent and Spain's 42.9 percent strongly supporting governmental action. The lowest percentile in this category was the Netherlands with only 21.2 percent expressing that their government should handle climate change as a very important issue. Interestingly, the 15.7 percent of survey respondents in the United States characterizing the reduction of climate change as unimportant politically is significantly higher compared to all other countries.

The mean values of each country suggest that countries can be grouped together based on how the public perceives the political saliency of climate change. As a result, the countries were divided into three groups. Group 1 consists of the countries Brazil and Mexico. With mean scores ranging from 3.34 to 3.47, the data show that the survey participants in these two countries take climate change very seriously and want their governments to be strongly involved. Spain, Germany, Japan, and Canada also seem to want their governments to place a high level of importance on reducing climate change. However, the means of these countries suggest that the participants do not want their government to put as much importance on climate change activities as compared to Brazil and Mexico. The mean scores of the second group range from 3.15 (Canada) to 3.26 (Spain). The third group consists of the three countries with a mean score below 3.00, United Kingdom, United States, and Netherlands. In addition, in all three countries, the percentage of people who do not perceive climate change as an important issue is above 20 percent.

This grouping of nine countries is supported by the Kruskal–Wallis test, which was performed to confirm that the means are significantly different among the nine country groups. The test suggests, with a significance level of $p < 0.01$, that there is a significant difference among the nine countries in terms of how the public perceives the political saliency of reducing *climate change*, thus indicating a relationship between the two variables "country of origin" and "level of importance for government to be involved in reducing *climate change*". However, the relationship between socioeconomic variables and climate change are discussed in the next chapter. The mean rating of each country confirms the validity of the country grouping. The mean rank of Brazil (4473.02) and Mexico (4153.34) are significantly different from the mean ratings of the countries belonging to Group 3, United Kingdom (3294.53), United States (3294.53), and Netherlands (2808.80). Japan's and Canada's mean scores are very similar, confirming the decisions to place them in the same group.

Political saliency for resolving climate change is also used as an indicator for the public's level of concern. Nonetheless, the survey asked specifically about the level of concern regarding possible climate change impacts in each of the nine countries. Responses were categorized on a 5-point scale from "not at all concerned" to "highly concerned". Data show that on the

Table 4.3 Public's level of concern regarding the possible impacts of climate change by country

	Not at All Concerned	Slightly Concerned	Somewhat Concerned	Concerned	Highly Concerned	Mean
Mexico	1.0%	0.8%	6.3%	35.0%	56.9%	**4.46**
Brazil	1.4%	1.8%	7.6%	28.5%	60.8%	**4.46**
Canada	4.3%	7.2%	17.8%	39.7%	31.0%	**3.86**
Spain	4.3%	7.7%	20.3%	39.5%	28.3%	**3.80**
Germany	4.4%	6.6%	25.4%	37.3%	26.5%	**3.75**
Japan	3.0%	20.7%	21.8%	26.2%	28.2%	**3.56**
United Kingdom	9.3%	15.6%	24.4%	31.4%	19.4%	**3.36**
United States	14.9%	11.8%	20.1%	30.5%	22.7%	**3.34**
Netherlands	10.4%	16.9%	28.9%	32.9%	11.0%	**3.17**

Note: Survey participants were asked on a scale from 1 to 5, where 1 is "not at all concerned" and 5 is "highly concerned", how concerned are you about the possible impacts of climate change?

international level, the majority of the 7,261 people participating in this study are concerned about climate change's potential adverse impacts. Over 80 percent of the survey participants stated at least some concern regarding the possible impacts of climate change and close to 32 percent indicated high levels of concern. Only 6.1 percent indicated that they are not concerned at all. As shown in Table 4.3, the data suggest that that Mexico with 56.9 percent and Brazil with 60.8 percent have the largest percentage of people who are highly concerned. In Mexico, 91.9 percent of the participants indicated that they are either "concerned" or "highly concerned", closely followed by Brazil with 89.3 percent. In Canada, Spain, and Germany the percentage of participants who are "concerned" or "highly concerned" ranges between 63.8 (Germany) and 67.8 percent (Spain). In the case of Japan, the percentage of people that are "slightly concerned" (20.7 percent) is larger compared to all other countries. In the United States, 14.9 percent of the participants state that they are "not at all concerned". This represents a significant difference compared to all other countries.

This leads to the question: why are people concerned about climate change? To gain insight into public attitudes, the survey asked respondents about three possible reasons – that climate change will be irreversible at some point; that there is a lack of political will to prevent climate change; and that they worry about the impacts on future generations more than the impacts in their life time. Responses were measured on a 5-point Likert scale ranging from "strongly disagree" to "strongly agree", and the results are illustrated in Table 4.4.

Table 4.4 Possible reasons for the public's concern about climate change

"I worry about climate change because at some point we will not be able to reverse it."

	Strongly Disagree	Disagree	Undecided	Agree	Strongly Agree
Mexico	1.7%	2.5%	5.6%	22.3%	67.9%
Brazil	1.8%	3.6%	7.4%	34.5%	52.8%
Spain	4.3%	3.4%	22.3%	36.4%	33.6%
Canada	5.8%	8.9%	20.6%	34.7%	30.1%
Germany	4.9%	6.8%	25.6%	39.2%	23.5%
Japan	2.1%	6.2%	30.5%	46.0%	15.3%
United States	13.5%	7.8%	21.5%	30.9%	26.2%
United Kingdom	8.3%	10.1%	28.3%	34.7%	18.5%
Netherlands	5.4%	18.4%	25.2%	35.0%	16.1%

"I worry about climate change because there is no strong political will to prevent it."

	Strongly Disagree	Disagree	Undecided	Agree	Strongly Agree
Mexico	2.7%	5.4%	11.1%	35.0%	45.8%
Brazil	2.4%	5.0%	10.0%	42.3%	40.4%
Spain	3.4%	6.6%	25.7%	38.9%	25.5%
Germany	3.9%	10.3%	25.5%	36.7%	23.7%
Japan	2.2%	7.2%	34.1%	43.1%	13.4%
Canada	5.8%	10.0%	26.7%	35.1%	22.4%
United Kingdom	7.9%	11.7%	35.1%	32.1%	13.1%
United States	14.4%	9.9%	28.3%	30.4%	17.0%
Netherlands	5.4%	22.5%	27.4%	33.1%	11.5%

"I do not worry much about climate change for me, but I worry for future generations."

	Strongly Disagree	Disagree	Undecided	Agree	Strongly Agree
Germany	4.5%	7.8%	22.2%	38.8%	26.7%
Canada	10.0%	18.4%	17.6%	35.6%	18.4%
Netherlands	7.4%	21.8%	18.9%	38.8%	13.0%
United Kingdom	8.4%	17.2%	28.6%	35.7%	10.1%
Brazil	17.0%	22.0%	9.9%	28.1%	23.0%
United States	14.4%	16.4%	23.4%	30.0%	15.8%
Japan	8.1%	22.6%	32.7%	27.7%	8.9%
Spain	17.8%	21.3%	26.6%	22.8%	11.6%
Mexico	27.7%	21.9%	11.6%	19.1%	19.6%

Note: Survey participants were asked on a scale from 1 to 5, where 1 is "strongly disagree" and 5 is "strongly agree", what is the level of agreement with the statements above?

Results show that there was large-scale agreement with the statement that at some point, climate change will not be able to be reversed. In total more than 75 percent of the participants agreed or strongly agreed with that sentiment. In Mexico 67.9 and in Brazil 52.8 percent of all participants strongly agreed with that statement. Almost 60 percent of all those surveyed agreed that their worry is based on the "lack of political will to prevent climate change". Similar to the question regarding political saliency, this statement was designed to measure how the public perceives their government's current level of involvement with the issue of climate change.

The frequency distribution for all nine countries shows that over 65 percent of the public either agrees or strongly agrees with the notion that the political will to address climate change is insufficient. Close to 21 percent are undecided and 13 percent of the survey participants either disagreed or even strongly disagreed (5.5 percent). Again, the percentage of people strongly agreeing with that attitude is the highest, by a significant margin, in Mexico (45.8 percent) and Brazil (40.4 percent), followed by Spain with 25.5 percent. The lack of political will as a reason to worry about climate change is most strongly contested among the populations of the Netherlands, the United Kingdom, and the United States. In these countries, at least 26 percent disagree or strongly disagree with the statement "I worry about climate change because there is no strong political will to prevent it". Comparing these results to the previously discussed political saliency question, it becomes apparent that populations that indicated that they want their government to act on climate change as important also believe that the political will to do so is deficient.

With respect to the "impacts on future generations factor", approximately 46 percent of the total sample indicated that they worry about future generations and not necessarily for themselves. The data suggest that the German population is the most worried about the intergenerational factor relative to the other countries with over 65 percent agreeing or strongly agreeing. Germany is followed by Canada with 54 percent, the Netherlands and Brazil both with 51 percent, the United States with 45 percent, Mexico with 38.7 percent, Japan with 37 percent, the United Kingdom with 36 percent, and Spain with 34 percent.

Public perceptions of climate change risks and threats

The survey instrument asked participants several questions regarding the risks and potential threats of a list of negative environmental events, such as climate change. The way the public perceives the risk and threats of climate change is important information for successful communication efforts. It allows risk communicators to better understand their audience in terms of their perceptions and experiences regarding climate change and its impact and develop effective communication tools. Thus, increasing the likelihood that climate change communication programs will enhance the public's level of awareness and sense of urgency as well as increasing their policy support through behavioral changes.

The first question in the survey instrument's risk and threat section addressed the perceived level of consequences or effects of different environmental changes expected over the next 20 years. One of the environmental changes listed was "Substantial increase in global warming resulting in global climate change". The participants determined the level of consequences of each environmental change based on a 5-point scale, ranging from "not likely at all to happen" to "serious negative consequences". The responses provided a first indication on whether or not the participants believed in the reality of climate change and how they perceive the level of consequences or relative risk compared to other environmental impacts. On the global scale, the data show that only 3.6 percent of the total global sample holds the belief that global warming will not result in climate change. However, over 55 percent of those surveyed indicated that climate change will have "very serious negative consequences"; another 24.9 percent expect "moderate negative consequences" over the next 20 years. Together, this percentage of the global survey represents a significant result related to peoples' beliefs about climate change and the level of negative consequences they expect on the global scale. In total, this question listed seven environmental changes for which responses in levels of perceived risk was asked for.

As shown in Table 4.5, with a mean score of 4.26 "substantial increase in global warming resulting in global climate change" only ranks fifth in terms of the public's perceived level of risk or consequence. A higher percentage of participants expect greater negative consequences from events such as the extensive loss of forests and/or wetlands, the deterioration of the ozone layer, the increasing frequency of droughts, and the rising number of major hurricanes and/or floods. The two environmental changes ranked 6th and 7th on the risk scale were "worsening of urban air pollution" and "further extinction of endangered animals and plants".

Table 4.5 Ranking of environmental changes based on the public's perceived level of risk

Rank	Environmental Changes	Mean
1	Extensive loss of forest and/or wetlands	**4.43**
2	Deterioration of ozone layer	**4.34**
3	Increasing frequency of droughts	**4.30**
4	Increasing frequency of major hurricanes and/or floods	**4.29**
5	Substantial increase in global warming resulting in global climate change	**4.26**
6	Worsening of urban air pollution	**4.24**
7	Further extinction of endangered animals and plants	**4.21**

Note: Survey participants were asked on a scale from 1 to 5, where 1 is "not likely at all to happen" and 5 is "serious negative consequences", over the next 20 years what do you think the level of consequences will be if any, of the following environmental changes?

As illustrated in Table 4.6, examining the frequency distribution for each country reveals significant national differences among the surveyed populations on the impacts of global warming. With the exception of the Netherlands, the United Kingdom, and the United States, the majority of the public in the remaining six countries expect serious negative consequences from increases in global warming resulting in climate change. Data indicate that the majority of populations in Mexico and Brazil perceive serious negative consequences if increases in global warming occur. On the other end of the spectrum, over 10 percent of US participants do not believe that climate change will happen at all. This suggests that the reality of climate change is most challenged by the US public than in any other surveyed country.

As mentioned before, Mexico and Brazil show a very similar frequency distribution and their mean values are also close to each other. Another group of countries with similar percentages and means consists of Spain, Japan, Germany, and Canada. Their means range from 4.27 to 4.35 indicating that the overall populations of these countries expect moderate to serious consequences. The third group includes countries that have a mean score below 4.0 such as the Netherlands, the United Kingdom, and the United States. For these countries the data suggest that the majority of people expect slight to moderate negative consequences from global climate change.

In the same segment of the questionnaire, another question asked participant's how they perceive the level of threat from climate change over the next 50 years. The previous question asked responders about their concern with a time period of 20 years. Participants were asked to determine the level of threat

Table 4.6 Perceived level of hazard consequences, if any, from future climate change

	Not Likely at All To Happen	No Negative Consequences	Slight Negative Consequences	Moderate Negative Consequences	Serious Negative Consequences
Mexico	0.2%	0.2%	1.9%	14.4%	83.2%
Brazil	1.4%	1.4%	3.5%	13.0%	80.8%
Spain	3.4%	2.4%	8.5%	27.0%	58.6%
Japan	1.1%	3.0%	11.6%	29.2%	55.1%
Germany	3.2%	1.6%	13.6%	27.4%	54.2%
Canada	2.4%	2.0%	13.4%	30.6%	51.6%
Netherlands	3.7%	6.0%	21.5%	28.1%	40.8%
United Kingdom	4.4%	5.6%	21.1%	29.7%	39.2%
United States	10.8%	6.4%	16.9%	26.3%	39.6%

Note: Survey participants were asked on a scale 1 from 5, where 1 is "not likely at all to happen" and 5 is "serious negative consequences", over the next 20 years what do you think the level of consequences will be if any, of a substantial increase in global warming resulting in climate change?

for four different groups; "plants and animals", "people in other countries", "people in your country" and "you and your family". The level of threat was measured using a 4 point Likert-scale ranging from 1 representing "no threat at all" to 4 being "a high threat".

Table 4.7 presents the results of a categorical index that was created based on how the public perceives the climate change level of threat over the next 50 years for "plants and animals", "people in other countries", "people in your country", and "you and your family" combined. The mean values and frequencies suggest that the public perceives climate change as a potential threat over the next 50 years. In terms of posing a high threat, the data show significant differences among the countries. In Mexico and Brazil, over 80 percent of respondents indicate that they believe climate change will be a high threat over the next 50 years. In contrast, only 19.2 percent in the Netherlands agreed with that sentiment. On the other end of the spectrum, over 15 percent of participants in the United Kingdom (15.2 percent), the United States (20.7 percent), and the Netherlands (22.9 percent) perceive climate change as not a threat or only a slight threat.

In total over 50 percent of the 7,261 survey participants recognize climate change as a high threat for plants and animals as well as for people living in other countries than themselves. However, when looking at the frequencies for individual areas of impact, "high levels of threat" are perceived predominantly for plants, animals, and people in other countries, and significantly lower for people in their respective countries, to their family, and to themselves. The data show that only 36.8 percent of surveyed population believes that climate change presents a high threat for other people in their country. In terms of the perceived level of threat for their family and themselves only 33.4 percent

Table 4.7 Perception of threat resulting from climate change over the next 50 years

	No Threat at All	A Slight Threat	Some Treat	A High Threat	Mean
Mexico	0.2%	0.8%	11.0%	87.9%	**3.87**
Brazil	0.3%	1.4%	14.3%	84.1%	**3.82**
Germany	1.8%	4.5%	33.5%	60.2%	**3.52**
Canada	1.5%	5.4%	32.8%	60.3%	**3.52**
Spain	2.7%	8.6%	32.0%	56.6%	**3.43**
Japan	1.0%	9.8%	42.5%	46.8%	**3.35**
United Kingdom	3.5%	11.7%	43.3%	41.5%	**3.23**
United States	8.2%	12.5%	30.0%	49.3%	**3.20**
Netherlands	2.8%	20.1%	58.0%	19.2%	**2.94**

Note: Under the assumption that global climate change continues to occur over the next 50 years, survey participants were asked on a scale from 1 to 4, where 1 is "no threat at all" and 4 is "very high threat", how would you rate the level of threat resulting from global climate change?

chose the "a high threat" response category. Only 9.5 percent of the surveyed publics feel that climate change does not pose any danger to themselves or their families over the next 50 years.

This finding of "distant risk" is further supported by the results of two other survey questions that asked participants to indicate how long it will take until dangerous climate change impacts will be experienced "somewhere on earth" and "in their region". Asking the public to determine a timeline for observable climate change impacts adds another perspective in terms of the perceived saliency of the issue and level of urgency among the population. The participants were asked to choose from six different answer categories: "impacts are already experienced", "in 10 years", "in 25 years", "in 50 years", "in 100 years", or "never". The results for each country individually are displayed in Table 4.8.

With the exception of the United Kingdom and the Netherlands, the data suggest that the majority of the nine-country public believes that climate change impacts are already being experienced somewhere on earth. In all nine countries, no answer category was chosen more often than the one stating that impacts are already being experienced. On the other hand, the answer category "never" received the least level of agreement from eight of the nine surveyed countries. In this case, the outlier is the United States where 14.3 percent of the participants indicated that they believe that no impacts will be experienced anywhere on the planet. When asked about a timeframe until climate change impacts will become apparent locally, a smaller percentage of people believe that dangerous impacts of climate change are occurring within their own region than somewhere on earth. A majority of over 70 percent in Brazil and Mexico believe that they are personally already experiencing climate change impacts. This statistic is much more than the country with the third highest-rating Germany, where 46.5 percent of the participants indicate that they already experience climate change impacts. Based on the total sample size, 42.8 percent of the nine-country population believes that dangerous impacts are being experienced today in the region they live in. In addition, just over 20 percent of the global surveyed population believes that they will experience effects from climate change within 10 years.

However, the data also suggest that most of the population in the United Kingdom does not believe that they are already experiencing climate change impacts. Only one out of five participants believe that climate change impacts are already occurring on the local scale. This is the lowest rate among all nine countries. Moreover, the United Kingdom, the Netherlands, and the United States are the only countries where over 50 percent of the surveyed populations believe that climate change impacts are neither already occurring locally nor will be in the next 10 years. In the remaining six countries, at least 60 percent of the public either seems to experience impacts of climate change already or at least expects them to take place locally within the next 10 years.

Overall, the results suggest that the overwhelming majority of the public either already experiences impacts or expects climate change impacts to occur globally and locally over the next 25 years. In total, 60.7 percent of

Table 4.8 Timeframe until the public believes global climate change impacts will be experienced

Somewhere on Earth

	Already Experienced	*In 10 Years*	*In 25 Years*	*In 50 Years*	*In 100 Years*	*Never*
Mexico	81.0%	9.4%	4.7%	3.1%	1.3%	0.4%
Brazil	83.0%	5.6%	5.9%	3.1%	1.9%	0.5%
Germany	65.5%	10.4%	11.3%	5.9%	3.5%	3.3%
Japan	59.7%	14.1%	12.5%	7.8%	3.7%	2.1%
Canada	64.4%	9.8%	10.2%	5.8%	6.3%	3.5%
Spain	53.0%	9.1%	14.3%	12.4%	7.6%	3.7%
United Kingdom	48.1%	10.4%	14.0%	11.0%	8.5%	8.0%
United States	50.5%	8.4%	12.5%	7.3%	7.1%	14.3%
Netherlands	45.0%	8.4%	13.3%	13.3%	11.8%	8.2%

In Your Region

	Already Experienced	*In 10 Years*	*In 25 Years*	*In 50 Years*	*In 100 Years*	*Never*
Mexico	72.0%	17.1%	5.8%	3.1%	1.5%	0.5%
Brazil	70.4%	15.1%	8.0%	4.0%	1.9%	0.6%
Germany	46.5%	22.0%	14.3%	8.7%	4.2%	4.2%
Japan	39.9%	24.5%	15.8%	9.3%	6.7%	3.9%
Canada	44.9%	16.4%	14.3%	10.8%	9.4%	4.1%
Spain	34.0%	26.7%	18.6%	11.8%	6.0%	2.9%
United Kingdom	34.7%	19.9%	13.7%	9.2%	7.4%	15.1%
United States	24.6%	20.3%	15.5%	15.1%	15.0%	9.5%
Netherlands	20.0%	20.6%	19.4%	17.3%	12.9%	9.8%

Note: The participants were asked to choose from six different answer categories: "impacts are already experienced", "in 10 years", "in 25 years", "in 50 years", "in 100 years", or "never" and state how long, if ever, it will take until dangerous impacts of global climate change will be experienced (a) somewhere on earth and (b) in their region.

the 7,261 survey participants hold the belief that dangerous impacts of climate change somewhere in the world are already being experienced today. In addition, 9.5 percent indicate that they expect impacts to be experienced somewhere on the planet within the next 10 years and another 11 percent thinks within 25 years. Still, in the United States, 15.3 percent and close to 10 percent of the populations in the United Kingdom and in the Netherlands do not think that they will ever experience any impacts of climate change in their own region.

Another principal question asked about the risk of global climate change possibly causing various negative impacts of climate change over 50 years. Respondents were given 13 consequences of climate change and were asked to evaluate their level of risk on a 5-point Likert scale. Table 4.9 shows the results for all nine countries combined. Among the possible consequences of climate change, for example are "more frequent and serious hurricanes", "coastal damage", "negative impacts on the global economy", or "more people living in poverty". Responses in the "very high" risk category show that the survey participants are most concerned about climate change causing "droughts and water shortage" (42.4 percent), "more frequent and serious floods (42.4 percent), "forest fires" (40 percent), and "severe heat waves" (39.7 percent).

Overall, at least 30 percent of the survey participants stated that there is a very high risk for climate change causing each of the 13 negative events

Table 4.9 Global trends for perceived level of risk over the next 50 years for climate change causing negative impacts

	No Risk	Little Risk	Moderate Risk	High Risk	Very High Risk
More frequent and serious hurricanes	2.7%	5.9%	23.7%	37.5%	30.2%
Greater extinction of plant and animal species	2.3%	5.7%	21.2%	33.7%	37.0%
Famines and food shortage	2.4%	5.3%	20.8%	34.3%	37.2%
Droughts and water shortages	2.1%	4.5%	17.0%	33.9%	42.4%
More people living in poverty	2.9%	5.6%	21.7%	34.4%	35.5%
More refugee problems in parts of the world	2.8%	6.6%	24.0%	35.1%	31.4%
Severe heat waves	2.4%	4.8%	19.3%	33.8%	39.7%
Forest fires	2.4%	4.8%	18.6%	34.2%	40.0%
Diseases/epidemics	2.8%	6.6%	23.2%	30.9%	36.5%
More frequent and serious floods	2.4%	4.1%	16.6%	34.5%	42.4%
Coastal damage	2.5%	5.2%	20.8%	34.7%	36.8%
Extensive loss of farmland	2.8%	5.9%	21.4%	33.3%	36.6%
Negative impacts of the global economy	3.1%	5.6%	21.7%	32.6%	37.1%

Note: Survey participants were asked on a scale from 1 to 5, where 1 is "no risk" and 5 is "very high risk", over the next 50 years, how do you rate the risk of global climate change causing ... ?

listed in Table 4.9. Moreover when the answer categories "high risk" and "very high risk" are combined, the data show not much variation between the risk perceptions of the different negative climate change impacts. Instead, between 66.5 percent and 76.9 percent of the surveyed population perceive the risk of climate change causing any of the listed impacts as high or very high. On the other hand, only a small percentage of the public seems to believe that there is no risk of global climate change causing any of the different negative events. For example, only 3.1 percent stated that they do not expect any negative impacts from climate change on the global economy in the near future. In addition, less than 3 percent of the participants do not anticipate an increase in poverty, refugee problems, or loss of farmland due to climate change. As a result, the data suggest that on average for all nine countries, 71.2 percent of the public perceives the risk of global climate change causing one of the 13 negative events as high or very high. Compared to only an average of approximately 8 percent who see "no to little risk" of climate change increasing the frequency or severity of environmental hazards.

On the national scale, the mean scores for the perceived level of risk of climate change causing any of the 13 environmental impacts suggests country-specific differences. As illustrated in Table 4.10 and based on the 5-point Likert scale ranging from no to very high risk (5 on the scale), the average mean scores were the highest for Mexico and Brazil. For these two countries, the average mean score was above 4.5, demonstrating that the populations from Mexico and Brazil perceive high risk levels of climate change. The lowest mean scores were from the United States, the Netherland, and the United Kingdom. The average mean score was below 3.7 indicating that a large number of people in these countries perceive climate change as a moderate risk for causing future environmental impacts.

Table 4.10 Mean scores by country for the public's perception of the risk of climate change causing various environmental impacts over the next 50 years

	BRA	CAN	GER	JP	MEX	NET	ESP	UK	USA
More frequent and serious hurricanes	4.28	3.91	4.00	3.79	4.53	3.54	3.92	3.51	3.42
Greater extinction of plant and animal species	4.49	4.02	3.97	3.83	4.57	3.59	4.05	3.76	3.59
Famines and food shortage	4.34	4.00	3.84	4.14	4.56	3.80	3.95	3.77	3.56
Droughts and water shortages	4.57	4.06	4.00	4.17	4.69	3.85	4.18	3.80	3.65
More people living in poverty	4.27	3.96	4.05	3.90	4.49	3.64	4.07	3.62	3.57

	BRA	CAN	GER	JP	MEX	NET	ESP	UK	USA
More refugee problems in parts of the world	4.14	3.91	4.00	3.77	4.27	3.64	3.85	3.66	3.56
Severe heat waves	4.62	4.13	4.05	3.98	4.66	3.64	4.08	3.63	3.66
Forest fires	4.64	4.08	4.02	3.91	4.62	3.77	4.19	3.65	3.62
Diseases/epidemics	4.47	3.99	3.79	4.04	4.61	3.57	3.93	3.50	3.48
More frequent and serious floods	4.55	4.14	4.14	4.02	4.70	3.90	4.18	3.82	3.58
Coastal damage	4.43	4.03	4.04	3.87	4.54	3.65	4.01	3.77	3.61
Extensive loss of farmland	4.39	4.03	3.92	4.03	4.64	3.48	4.02	3.56	3.59
Negative impacts of the global economy	4.40	4.08	3.88	3.98	4.55	3.51	4.03	3.59	3.67
AVERAGE	**4.43**	**4.03**	**3.98**	**3.96**	**4.57**	**3.66**	**4.04**	**3.66**	**3.58**

Note: Survey participants were asked on a scale from 1 to 5, where 1 is "no risk" and 5 is "very high risk", over the next 50 years, how do you rate the risk of global climate change causing ...?

The public's perceived knowledge, belief, and attitude towards climate change

The literature shows that when people have a better understanding of climate change science, they tend to be more supportive of mitigation efforts (Read et al., 1994). Table 4.11 shows the frequency distribution for all countries combined for their perceived level of knowledge regarding different aspects of climate change. Less than 60 percent of the public feels "informed" or "very informed" about the impacts, causes, and ways to reduce climate change. On a 1 to 4 scale, where 1 is "not informed" and 4 is "very informed", the public's perceived knowledge averages between 2.0 and 3.0. Moreover, less than one-third (23.3 percent) of the surveyed population feel at least informed about the various national and international policies to prevent climate change. Another 22.7 percent indicated that they do not feel informed at all about existing climate change policies. Also noteworthy are the fairly high numbers of people who indicated that they only are somewhat informed about these four important aspects of climate change. This indicates high levels of uncertainty among the public which can lead or enforce already existing misconceptions.

The public's preference towards four general climate change strategies and the level of belief in the reality of global climate change were used to describe a person's attitude towards. For the first attitude variable, the survey respondents were asked to choose one out of four possible general strategies on global climate change that comes closest to their opinion. These strategies and the frequency distribution by country are displayed in Table 4.12. By asking the

Table 4.11 Global trends for perceived level of knowledge about key aspects of climate change

	Not Informed	Somewhat Informed	Informed	Very Informed	Mean
The causes of global climate change	7.1%	36.7%	41.0%	15.2%	**2.64**
The impacts of global climate change	6.3%	33.9%	43.1%	16.7%	**2.70**
The ways in which we can reduce global climate change	10.5%	37.7%	37.7%	14.2%	**2.56**
The various national and international policies to prevent global climate change	22.7%	46.5%	23.3%	7.5%	**2.16**

Note: Survey participants were asked on a scale from 1 to 4, where 1 is "not informed" and 4 is "very informed", how well informed do you think you are about . . . ?

Table 4.12 Public's attitude towards taking political action against climate change

a) We should not take any steps that would have economic costs until we are certain that global climate change is really a problem.

BRA	CAN	GER	JP	MEX	NET	ESP	UK	USA
1.4%	11.9%	6.8%	5.1%	1.2%	12.2%	7.1%	14.2%	23.5%

b) We should take some steps just in case global climate change is real.

BRA	CAN	GER	JP	MEX	NET	ESP	UK	USA
9.0%	19.3%	10.7%	51.5%	4.7%	25.6%	16.1%	28.8%	22.0%

c) We only should take steps to address global climate change that are low in costs.

BRA	CAN	GER	JP	MEX	NET	ESP	UK	USA
1.6%	8.2%	14.3%	3.9%	4.1%	14.7%	7.2%	13.7%	10.7%

d) Global climate change is a serious problem, and we should begin taking steps now even if this involves significant costs.

BRA	CAN	GER	JP	MEX	NET	ESP	UK	USA
88.0%	60.7%	68.2%	39.6%	90.0%	47.5%	69.7%	43.3%	43.8%

Note: Survey participants were asked to choose one out of four possible general strategies on global climate change that comes closest to their opinion.

participants to choose one particular basic strategy, the person's attitude in terms of the general long-term policy approach become apparent. Together with the public's level of belief in the reality of climate change it, is possible to draw conclusions regarding someone's general attitude towards climate change.

Table 4.13 Public's level of belief in the reality of climate change

	Strong Believer	Believer	Moderate Believer	Non-Believer
Mexico	98.5%	1.1%	0.4%	0.0%
Brazil	98.4%	1.1%	0.1%	0.4%
Japan	95.3%	2.7%	0.8%	1.2%
Spain	94.0%	2.3%	1.2%	2.4%
Germany	93.6%	3.0%	1.3%	2.1%
Canada	93.5%	3.2%	1.5%	1.9%
United Kingdom	86.7%	4.9%	3.1%	5.3%
Netherlands	85.0%	6.9%	3.1%	5.0%
United States	80.7%	4.6%	4.4%	10.2%

Note: Result of an index created from the responses of several questions, based on level of belief in the reality of climate change.

With the exception of Japan, the strategy chosen most often by the participants was that "climate change is a serious problem and we should begin taking steps now even if this involves significant costs".

With regards to Japan, the data indicate that the majority of the population prefers the option of taking "some steps just in case climate change is real". However, when combining two answer categories, the data generally indicate that a significant number of people in Canada (21.1percent), Germany (21.3 percent), the Netherlands (26.9 percent), the United Kingdom (27.9 percent), and in the United States (34.3 percent) either oppose any policies that might hurt the economy or only support policies that are low in costs. The second attitudinal variable is an index that was created from the responses of several questions, based on level of belief in the reality of global climate change. As shown in Table 4.13, the overwhelming majority of the population strongly believes that climate change is real. Nevertheless, a small percentage of 10.2 percent in the United States are still not convinced that climate change is occurring.

The public's perceived levels of trust towards climate change information and levels of responsibility to reduce climate change

Besides building trust through interpersonal relations or ideological values and norms, people can also hold trust in organizations and institutions (Hardin, 2006). The role of trust is an important perceptual dimension that influences the success of policies targeting climate change as well as the public's willingness to commit to behavioral changes. We know from prior studies that when people have a better understanding of climate change science and trust in the information, they tend to be more supportive of mitigation efforts (Read et al., 1994; Bord et al., 1998). Furthermore, research shows that the failures of risk communication are significantly influenced by the public's trust in the communicator and in the ability of certain individuals, industries, or institutions responsible for risk management (Renn & Levine, 1991; Kasperson et al., 1992; Nye Jr. et al., 1997). If there is no trust in the source, any message and policies are likely to be disregarded, no matter how well designed and well delivered. Thus, public trust in organizations whose risk management addresses adaptation and mitigation strategies is vital in order to generate social cooperation to increase their likelihood of success. Therefore, the survey questionnaire asked about the public's level of trust towards different sources of information that also play a role in the design, communication, and/or implementation of climate change policies.

The question was measured on a 5-point Likert scale and asked: "On a scale of 1 to 5, where 1 is 'strongly distrust' and 5 is 'strongly trust' what is your level of trust towards the following sources of information regarding global climate change?" As shown in Table 4.14, the question included nine different sources of information. Data show that private sector corporations, governmental organizations, and religious leaders the least trusted as climate change information sources – less than 5 percent of the surveyed population expressed trust towards any of them. The majority of the population trust scientists the most, with 66.8 percent "somewhat or strongly" trusting scientists, followed by television weather reports (61.3 percent), family and friends (56.5 percent), and environmental organizations (56.1 percent).

Table 4.14 Global trends for public's level of trust towards different sources of climate change information

	Strongly Distrust	Somewhat Distrust	Undecided	Somewhat Trust	Strongly Trust	Mean
Television weather reports	5.4%	13.8%	19.4%	51.6%	9.7%	**3.46**
Corporations	14.1%	24.9%	36.0%	22.0%	3.0%	**2.75**
Family and friends	2.6%	7.8%	33.1%	41.6%	14.9%	**3.58**
Governmental organizations	16.0%	25.6%	28.8%	25.9%	3.7%	**2.76**
Environmental organizations	8.9%	12.5%	22.5%	39.1%	17.0%	**3.43**
Mainstream news media	9.4%	18.1%	30.5%	36.0%	6.0%	**3.11**
Scientists	3.4%	8.1%	21.7%	45.0%	21.8%	**3.74**
Religious leaders	33.3%	21.3%	27.3%	13.6%	4.6%	**2.35**
Teachers	7.8%	13.7%	40.0%	32.2%	6.3%	**3.16**

Note: Survey participants were asked on a scale from 1 to 5, where 1 is "strongly distrust" and 5 is "strongly trust", what is your level of trust towards the following sources of information regarding global climate change?

However, this study identified several contradictions and hesitations that indicate major uncertainties across all countries, not just in the United States as identified in previous studies (O'Connor et al., 1999; Moser, 2006; Leiserowitz et al., 2010). Survey participants were asked attitudinal questions, which tested their level of agreement with two different statements related to trust in the available scientific data and knowledge of climate change. Table 4.15 shows the two statements and provides data on the percentage of people for each of the nine countries who agreed or strongly agreed with them.

The analysis indicates a contradiction between the level of trust towards the scientists and the level of agreement with the sufficiency and trustworthiness of scientific data and expert knowledge. Although 66.8 percent of the total sample "somewhat or strongly" trusts scientists as climate change information sources, the data also indicate that many people doubt that the scientific community actually has enough data to fully understand the complexity of the issue.

The populations of Japan and Germany, in particular, seem to believe that more research needs to be conducted. In both countries, less than 20 percent of the participants agreed with the statement that the existing body of knowledge is sufficient. In fact, 51 percent in Japan and 49 percent in Germany specifically disagreed with that attitude, more than in any other of the countries. Furthermore, less than 50 percent of the public in the remaining seven countries trusts existing scientific findings. Only in Mexico and Brazil did more than 50 percent of survey respondents consider the scientific findings as trustworthy. This finding is significant and may help explain the difficulties in reaching a global consensus on reducing GHG emissions.

Uncertainties remain among populations when it comes to climate change science, however. Results indicate that a large number of people doubt the validity of the existing body of knowledge. This uncertainty is emphasized by the high percentage of respondents who indicated that they are undecided on whether or not to trust any of the sources of climate change information listed in the survey instrument. "Undecided" responses were documented for

Table 4.15 Strong to very strong level of trust in global climate change science

"We already have enough scientific data and expert knowledge to fully understand all aspects of global climate change."

BRA	CAN	GER	JP	MEX	NET	ESP	UK	USA
55.7%	36.2%	19.9%	15.4%	31%	38.3%	39.3%	26.4%	32.6%

"The scientific findings on global climate change is trustworthy."

BRA	CAN	GER	JP	MEX	NET	ESP	UK	USA
63.9%	47.3%	31.8%	32.1%	51.7%	26.2%	41.7%	33.2%	39.1%

Note: Survey participants were asked on a scale from 1 to 5, where 1 is "strongly disagree" and 5 is "strongly agree", what their level of agreement is with different statements related to trust in global climate change science, adaption and mitigation policies, and role of the media? Results shown in the table are the percentages of people who agreed or strongly agreed with the two statements illustrated.

teachers (40 percent), corporations (36 percent), family and friends (33.1 percent), and the mainstream news media (30.5 percent). Results also show that while the public perceives climate change as an issue that has to be solved to a large degree by the government and other institutions, governmental organizations rank among the least trusted climate change information sources.

Table 4.16. shows how people perceive responsibilities of different sectors to reduce climate change. For many hazards the public feels that applicable strategies exist in which they can be engaged in (Rotter, 1966). However, climate change is characterized by high uncertainties, unfamiliar risks, and other characteristics of hazards that make personal connections, responsibility, and engagement difficult. Many hope that effective communication efforts can foster a personal connection to climate change, raise the level of concern, and thus increase the level of support for mitigation and adaptation policies. The data show that the majority of people feel personal responsibility to take action to reduce the causes of climate change.

For many hazards, the public feels that applicable strategies exist in which they can be engaged in (Rotter, 1966). However, climate change is characterized by high uncertainties, unfamiliar risks, and other characteristics of hazards which make personal connections, responsibility and engagement difficult. Many hope that effective communication efforts can foster a personal connection to climate change, raise the level of concern, and thus increase the level of support for mitigation and adaptation policies. The data show that the majority

Table 4.16 Public's perceived level of responsibility towards different groups to mitigate the main causes of climate change

	Not Responsible	Somewhat Responsible	Responsible	Very Responsible	Mean
National/Federal Government	6.7%	23.5%	31.8%	38.0%	**3.01**
Environmental Protection Agency	6.8%	22.8%	35.0%	35.4%	**2.99**
The United Nations	6.7%	23.1%	32.7%	37.5%	**3.01**
Environmental Groups	10.6%	26.6%	34.8%	27.9%	**2.80**
Corporations	7.2%	25.5%	33.7%	33.7%	**2.94**
State Government	7.8%	26.1%	34.0%	32.1%	**2.90**
Local Government	8.8%	28.1%	33.9%	29.2%	**2.84**
Your Community	8.5%	29.7%	33.8%	28.1%	**2.81**
You Personally	8.0%	29.3%	33.5%	29.2%	**2.84**

Note: Survey participants were asked on a scale from 1 to 4, where 1 is "not responsible" and 4 is "very responsible", how responsible do you feel each of the following is for reducing the main causes of global climate change?

of people feel personal responsibility to take action to reduce the causes of climate change.

However, the mean score identifies several groups, agencies, or institutions for which the public believes that are more responsible than themselves to mitigate climate change. Although, differences are very small, on the global scale the public seems to perceive the national or federal government the primary institution responsible to reduce climate change, followed by the United Nations, the country's environmental protection agency, and corporations. Among the nine possible groups responsible for reducing climate change, personal responsibility ranks sixth. Thus, the public seems to perceive climate change as an issue that has to be solved to large part by the government and other institutions. This poses a significant barrier to the success of mitigation and adaptation strategies, since many of these policies require the public's support and cooperation.

Public's support for climate change policies and willingness to commit to behavioral changes

As discussed in the introduction of this book, a multitude of different planning approaches and climate change policies already exist to mitigate and adapt to climate change. However, without the public support, these strategies will not be successful nor will decision makers have the political capital to implement them in the first place. Many climate change policies and strategies need to be supported by the public though often additional financial burdens or behavioral changes. As a result the survey questionnaire included several questions to determine the public's attitudes towards various national climate change policies. Past studies show that the public largely supports policies that impact climate change at the national and international scale in general, but resists tax policies that directly support it especially in the United States (O'Connor et al., 1999; Moser, 2006; Leiserowitz et al., 2010). This study supports these findings on a global scale as well as for different countries. The data show that less than a quarter of the total sample supports tax hikes as economic incentives to reduce the use of electricity or automobiles. No more than one-third of the public supports tax hikes for mitigation policies.

The participants were asked on a 5-point Likert scale how much they support or oppose each of several policies. The 5-point Likert scale consists of the answers "strongly oppose", "moderately oppose", "undecided", "moderately support", and "strongly support". Additive indexes were created to aggregate the different strategies into the two different groups, "Overall support for mitigation policies" and "Overall support for adaptation policies", and to get a better sense of the public's general support for climate change mitigation and adaption in the nine different countries. Mitigation addresses the core cause of human induced climate change namely the large amount of energy consumption and the resulting of greenhouse gas emissions. The concept of mitigation is clearly understood by scientists and decision makers. Adaptation strategies focus on avoiding negative impacts caused by global climate change. They are essentially adjustments with the aim to increase resilience or decrease vulnerability to current or expected impacts of climate change. The indexes were calculated based on the degree the participants supported the following policies:

- Index 1: Overall support for mitigation policies
    - o Require higher fuel efficiency for automobiles
    - o Require higher energy efficiency standards for buildings, household appliances, material production, and building methods
    - o Require higher taxes on electricity
    - o Require electric utilities to produce at least 20 percent of their electricity from renewable energy sources by the year 2020
    - o Provide subsidies to industries to invest in alternative energy development
    - o Require higher road taxes and tolls
    - o Require installation of solar panels or photovoltaics on buildings
    - o Require more compacts, higher density, mixed use, and transit oriented development

- Index 2: Overall support for adaptation policies
    - o Require cities over the next 220 years to invest in coastal flood protection and barriers
    - o The national/federal government should mandate that I personally take action to respond to undesirable impacts of climate change
    - o The national/federal government should mandate that local governments take action to respond to undesirable impacts of climate change
    - o The national/federal government should encourage action to respond to undesirable impacts of climate change
    - o The national/federal government should make me aware of how climate change may affect me

Compared to mitigation, adaptation is a local challenge since climate change impacts can vary between regions. As a result, the adaption questions were designed in a more general way to ensure they can be answered by people in different countries facing different impacts.

Table 4.17 shows the frequency distribution for the complete survey sample of the nine countries combined and indicates the global trends in terms of public support for climate change policies and strategies. Responses show that only a small percentage opposes any effort to act on the sources or impacts of climate change. Only 7.2 percent of all participants oppose mitigation policies, 35.3 percent are undecided, and the majority, 57.5 percent, moderately to strongly support mitigation policies in general. On the national scale, the Dutch and the United States moderately or strongly oppose mitigation policies by about 16 percent each. In the remaining seven countries, the percentage of people opposing policies to address the causes of climate change is less than

Table 4.17 Global trends for public's level of support for climate change policies

	Strongly Oppose	Moderately Oppose	Undecided	Moderately Support	Strongly Support	Mean
Overall support for mitigation policies	2.2%	5.0%	35.3%	42.6%	14.9%	**3.63**
Overall support for adaptation policies	3.1%	4.1%	26.8%	40.4%	25.6%	**3.81**

Note: Survey participants were asked on a scale from 1 to 5, where 1 is "strongly oppose" and 5 is "strongly support", how much do you support or oppose the following policies to reduce global climate change? The indexes were calculated based on the degree the participants supported the different mitigation and adaption policies.

8 percent. At least 45 percent of the public among all nine countries generally support mitigation policies. In Germany, Japan, the Netherlands, and the United Kingdom, 40 percent or more of participants indicated they are undecided whether they should support mitigation policies or not. In the remaining countries, the undecided percentage is less but still substantial, ranging from 21 percent in Mexico to 36 percent in the United States.

The response pattern for adaptation is similar to mitigation, both on the global and national levels. When combining all survey participants, only 7.2 percent oppose adaptation strategies, 26.8 percent are undecided, and 65 percent at least moderately support the idea that adapting to climate change through policies and strategies is necessary. There is, however, relatively strong opposition to climate change adaptation in the United States compared to all other countries. About 20 percent of US respondents indicated that they would not support adaptation policies. In addition, the United States and the United Kingdom are the only nations where fewer than 50 percent of participants seem to moderately or strongly support adaptation policies. The results show ambiguity between the non-binding relatively strong support for adaptation and mitigation policies in general and the public's support for specific policies, willingness to pay more money for climate change abatement, and willingness to engage in behavioral changes to reduce climate change.

In terms of the public's willingness to change its behavior (such as in support for mitigation policies), the data show that most people are willing or even strongly willing to do so. Table 4.18 shows the frequency distribution for the six behavioral options for all nine countries combined. With close to 80 percent of the participants indicating a willingness or even strong willingness, the survey results demonstrate that the public is most willing to use more recycled paper and purchase energy saving appliances. However, on the national scale, only 25.1 percent of the participants from Japan indicated a strong willingness to mainly use recycled paper. This is significantly less compared to the top two countries Mexico with 68.5 percent and Canada with 54.7. Japan also ranked last for the three behavioral questions addressing energy consumption at

Table 4.18 Global trends for public's level of willingness to change their behavior

	Not Willing at All	Slightly Willing	Undecided	Willing	Strongly Willing
Use public transit for most of my travel	15.1%	15.7%	14.3%	29.4%	25.5%
Install solar panels	9.8%	10.0%	17.3%	33.8%	29.0%
Buy mainly locally produced goods	5.3%	9.3%	16.0%	39.5%	29.8%
Use mainly recycled paper	3.6%	6.7%	10.0%	38.4%	41.3%
Purchase only energy saving appliance	3.3%	6.3%	12.3%	37.1%	40.9%
Insulate your home or apartment	5.6%	6.5%	18.3%	36.3%	33.2%

Note: Survey participants were asked on a scale from 1 to 5, where 1 is "not willing at all" and 5 is "strongly willing", how much are you willing to change your behavior in the following areas to reduce the causes of global climate change?

home. When asked about the level of willingness to purchase only energy saving appliances, install solar panels, or insulate their home or apartment, less than 11 percent of the survey participant in Japan answered with "strongly willing".

Overall the data indicate that the public is the least willing to change their travel behavior. Over 30 percent of the survey participants stated that they are not willing or only slightly willing to increase their use of public transit systems. In the United States and the Netherlands, the amount of people stating that they are not willing at all to use public transit for most of their travel is significantly higher compared to the other seven surveyed countries. In the United States, over 33 percent and in the Netherlands almost 25 percent did not show any willingness to use public transit more often. In addition, a relatively large number of people seem to be undecided in terms of their willingness to install solar panels (17.3 percent), buy mainly locally produced goods (16 percent), and improve the insulation of their homes or apartments (18.3 percent). The mean scores suggest that people are most willing to change their behavior in areas that do not impact their daily routine or cost extra money such as using mainly recycled paper and purchasing only energy saving appliances. On the other hand, insulating the home and installing solar panels will save money in the long run but requires an upfront capital investment first. Changing travel habits is a significant change of someone's daily routine and locally produced goods are often more expansive than mass produced products sold by the big-box supermarkets.

The behavioral questions listed in the previous table were also combined into an index capturing the public's overall willingness to change its behavior to reduce the causes of climate change (mitigation). The results of this index for each country are shown in Table 4.19. Based on the country-specific frequency

Table 4.19 Public's overall willingness to change their behavior to reduce the causes of climate change

	Not Willing at All	Slightly Willing	Undecided	Willing	Strongly Willing	Mean
Mexico	0.4%	0.6%	2.3%	31.1%	65.6%	**4.61**
Brazil	0.1%	2.4%	5.4%	33.9%	58.3%	**4.48**
Canada	0.9%	2.2%	11.1%	34.9%	50.8%	**4.32**
Spain	1.5%	2.8%	13.0%	39.2%	43.5%	**4.20**
Germany	2.9%	4.1%	17.5%	39.3%	36.2%	**4.02**
United Kingdom	2.1%	5.7%	15.2%	42.3%	34.7%	**4.02**
Netherlands	2.0%	6.6%	19.2%	44.6%	27.7%	**3.89**
United States	4.0%	7.0%	16.6%	41.5%	30.9%	**3.88**
Japan	0.7%	8.1%	27.7%	45.4%	18.1%	**3.72**

Note: Survey participants were asked on a scale from 1 to 5, where 1 is "not willing at all" and 5 is "strongly willing", how much are you willing to change your behavior in the following areas to reduce the causes of global climate change? The indexes were calculated based on the degree the participants indicated their willingness to commit to behavioral changes.

distribution and mean scores, the populations of Mexico and Brazil are the most willing to commit to behavioral change. Among the nine countries, the survey results for Japan show the least amount willing to change their behavior (63.5 percent), but the highest numbers for people that are undecided (27.7 percent). In terms of similarities, the mean scores and frequency distributions of Germany and the United Kingdom suggest similar behavioral attitudes among the populations of both countries. In Germany 75.5 percent and in the United Kingdom 77 percent of the survey respondents seem to be willing or strongly willing to make behavioral changes. In general the population of Mexico, Brazil, and Canada seem to be the most willing to adjust their behavior in order to mitigate climate change.

Summary

One of the underlying hypothesis of the basic frequency analysis was that the public perception of climate change in terms of threat and risk, saliency of the issue, trust in climate change information, and acceptable public strategies vary among countries. The data do confirm differences between some countries but also show similarities between others. The populations of Mexico and Brazil seem to be the most concerned about climate change, perceive it as a high risk, and want their respective governments to take stronger action against the impacts and causes of climate change. Whereas the survey participants of the Netherlands, United Kingdom, and the United States always were among the countries with the lowest amount of concern for climate change impacts, threats, and risks. Data also show that more survey participants in these three countries challenge the reality and/or danger of climate change than in any other country. A third group of countries consisting of Japan, Canada, Germany, and Spain also showed similar frequency distributions in regards to the political saliency, risks, and threats of climate change. According to the data, the public of these four countries are not as concerned as people living in Brazil and Mexico, but still perceives climate change as a significant issue, supports government involvement, and only a very small percentage of people doubts the existence of climate change and its potential negative impacts. Despite difference between the nine countries in terms of the perceived political saliency and the risks and threats of climate change, the results show similar trends among all nine countries. For example, in all nine countries, a significant amount of participants indicated that they worry about climate change, are concerned about its possible impacts, and perceive it as a politically salient issue.

Furthermore, a substantial percentage of people believe climate change impacts are already or soon will be experienced somewhere on earth. Simultaneously fewer people, but still a substantial percentage, believe that their own regions are experiencing dangerous impacts or will within the next 10 years. Several survey questions addressed the public's perceived level of threat from global climate change. The data support the argument that the lay public perceives climate change as a future threat. The results of the analysis also indicate significant differences in the perceived level of threat among the nine countries. About 85 percent of the respondents from Mexico and Brazil labeled climate change as a "high threat", more often than any other surveyed national population. In contrast, Japan, the United Kingdom, United States, and the Netherlands less than 50 percent of their populations perceive climate change as a high threat. About 85 percent of the respondents from Mexico and Brazil labeled climate change as a "high threat", more often than any other surveyed national population. In contrast, in Japan, the United Kingdom, United States, and the Netherlands, less than 50 percent of their populations perceive climate change as a high threat. Moreover, people who already have experienced climate change or believe they will experience climate change soon are more concerned about it compared to people who believe they will not experience impacts from climate change in the future.

In regards to the public's level of trust towards different sources of information, scientists seem to be most trusted in all nine countries, followed by family and friends. On the contrary, the public in all nine countries seem to share high levels of distrust towards religious leaders, governmental organizations, and corporations. Another significant finding is that the high percentages of people in all nine countries who indicated that they are undecided or uncertain towards all the listed sources of information. This indicates a high level of general uncertainty towards the issue of climate change. This notion is further supported by seemingly contradicting survey results from similar questions or rather statements. Although scientists are highly trusted and the most trusted source in all nine countries results also indicates a significant amount of people doubting the validity and sufficiency of the existing body of knowledge. This contradiction is another indication for the uncertainty the public seems to experience when being confronted with the issue of climate change.

The last segment of questions discussed in this chapter focused on the public's support for adaptation and mitigation policies in general without looking at specific strategies. Based on the data, the majority of the public seems to support efforts to reduce the causes and impacts of climate change. In fact, among the nine countries, the Netherlands was the only country where less than 50 percent, but still a significant amount, supported mitigation policies. However, similar to the public's level of trust towards sources of climate change information, a large number of people indicated that they are undecided to whether or not support any climate change strategy. With climate change being still a controversial topic in the political arena and among some groups of the population, the people who are undecided today could make the difference in the future success of various climate change policies. Finally, while the mean scores suggest that people are most willing to change their behavior in areas that do not significantly impact their daily routine or cost extra money (i.e. purchasing recycled paper and energy saving appliances), there is an opportunity for both communication and policy development. The overall analysis shows that more than 50 percent of respondents are willing to change their behavior in all categories. As shown in Table 4.5, more than 50 percent indicated they are willing or are strongly willing to change their travel behavior and use public transit systems more often. There is still some hesitancy related to behavior change, however, as indicated by the number of undecided responses in terms of their willingness to install solar panels (17.3 percent), buy mainly locally produced goods (16 percent), and improve the insulation of their homes or apartments (18.3 percent). Therefore, it is crucial to gain a better understanding what factors influence the public in their decision process.

Bibliography

Bord, R.J., Fisher, A., & O'Connor, R.E. (1998). Public perceptions of global warming: United States and international perspectives. *Climate Research, 11*, 75–84.

Hardin, R. (2006). *Trust*. Cambridge, UK: Policy Press.

Kasperson, R., Golding, D., & Tuler, S. (1992). Social distrust as a factor in sitting hazardous facilities and communicating risks. *Journal of Social Issues, 48*(4), 161–187.

Leiserowitz, A.A. (2005). American risk perceptions: Is climate change dangerous? *Risk Analysis, 25*(6), 1433–1442.

Leiserowitz, A., Maibach, E., & Roser-Renouf, C. (2010). *Climate Change in the American Mind: Americans' Global Warming Beliefs and Attitudes in January 2010*. New Haven, CT: Yale University and Mason University.

Moser, S.C. (2006). Talk of the city: Engaging urbanities on climate change. *Environmental Research Letters, 1*, 1–10.

Nye Jr., J.S., Zelikow, P.D., & King, D.C. (1997). *Why People Don't Trust Government*. Cambridge, MA: Harvard University Press.

Ockwell, D., Whitmarsh, L., & O'Neill, S. (2009). Reorienting climate change communication for effective mitigation: Forcing people to be green or fostering grass-roots engagement? *Science Communication, 30*(3), 305–327.

O'Connor, R., Bard, R., & Fisher, A. (1999). Risk perceptions, general environmental beliefs, and willingness to address climate change. *Risk Analysis, 19*(3), 461–471.

Read, D., Bostrom, A., Granger Morgan, M., Fischhoff, B., & Smuts, T. (1994). What do people know about global climate change? 2. Survey Studies of educated laypeople. *Risk Analysis, 14*(6), 971–982.

Renn, O., & Levine, D. (1991). Credibility and trust in risk communication. In R.E. Kasperson and P.J.M. Stallen (Eds.), *Communicating Risks to the Public* (pp. 175–218). Dordrecht, The Netherlands: Kluwer Academic.

Rotter, J. (1966). Generalized expectancies for internal versus external control of reinforcements. *Psychological Monographs, 80*, 1–28.

5 Relationships between perception factors, socioeconomic characteristics, and climate change policy support

Chapter 5 presents the results of the advanced statistical analysis of the survey data through crosstabs, standard multiple regressions, and stepwise regressions, and it addresses the remaining three research questions of the "Global Survey on Public Attitudes towards Climate Change" research project:

- What importance do climate change risk perceptions and attitudes play in the public's willingness to support mitigation and adaptation strategies?
- How do the public perceptions regarding climate change and attitudes towards mitigation and adaptation strategies vary by socioeconomic factors?
- What role do levels of knowledge and perceptions of trust and responsibility play in the public's level of support for adaptation and mitigation policies?

Crosstabs were used to explore relationships between different categorical indexes. Similar to the analysis of frequencies, the crosstabs also provide helpful insights to interpret the results of different regression analyses. For each crosstab, two tests were performed to ensure statistical significant relationships between the variables and between particular cells of the crosstab table. In particular, the chi square test is used to determine if there is a relationship between the two categorical variables. This study only considered a relationship between variables if the chi square test resulted in at least $p < 0.05$ because the value of 0.05 is the conventionally considered threshold of statistical significance (Field, 2009).

Standard linear regressions were used, for example, to explicate the relationships between independent variables such as risk perceptions, attitudes, and socioeconomic characteristics and dependent variables such as the public's level of support for climate change policies and their willingness to commit to behavioral changes. Regression analysis not only allows the confirmation of relationship between predictor variables (independent variable) and an outcome variable (dependent variable), but also enables the determination of the strength of the relationship and the amount of variability in one variable that is shared by the other.

Three outputs of the regression analysis are important in this research. First, the multiple correlation coefficient (R) measures the strength of the correlation between the predictor variables and the outcome variables. In general, an R score of 0.5 and higher indicates that the independent variables have strong effects on the dependent variables, whereas a value of less than 0.3 suggests a weak relationship (Field, 2009). Second, the coefficient of determination (R^2) illustrates how much the independent variables can explain variation in the dependent variables. Since multiple independent variables per regression can raise the R^2 and be a potential source for error, this study reports on the adjusted R^2 in cases with more than two predictor variables in a regression model. The adjusted R^2 compensates for the use of more predictors and adjusts the value downward (Field, 2009). Third, the analysis of variance (ANOVA) test has two important purposes. First, ANOVA test was used to determine whether or not a regression model predicts an outcome variable well and furthermore, confirms that the results are statistically significant and can be generalized for the countries' entire population. This was considered the case when the calculated F-ratio was significant at $p > 0.001$, which means that there is less than a 0.1 percent chance that the particular F-ratio would happen if the null hypothesis were true.

Stepwise regressions were conducted in cases where the standard multiple regressions showed a large effect between predictor variables and outcome variables. The aim of the stepwise regression was to determine the subset of independent variables that have the strongest relationship to a dependent variable. In stepwise regressions, the predictor variables are entered into the model based on their statistical contribution in explaining the variance in the dependent variable. First, the predictor that has the highest simple correlation with the outcome variable is entered into the model. If this predictor significantly improves the ability of the model to predict the outcome, then this predictor is retained in the model and the computer searches for a second predictor. The criterion used for selecting this second predictor is that it is the variable that explains most of the remaining variation of the outcome variable. Each time a predictor is added to the equation, a removal test is made of the least useful predictor, thus identifying the single independent variable that has the strongest relationship to the dependent variable.

Risk perception, attitudes, and support for climate change policies

The following discussion explores the role climate change risk perceptions and attitudes play in the public's willingness to support mitigation and adaptation policies, and thus addresses the following research question of the "Global Survey on Public Attitudes towards Climate Change" research project:

•   What importance do climate change risk perceptions and attitudes play in the public's willingness to support mitigation and adaptation policies?

In this phase of the analysis, two hypotheses were tested to structure the analysis and to discover the insights necessary to answer the research question. The hypotheses were tested through frequency distributions, crosstabs, and regression analyses and are as follows:

- The public's general support for mitigation and adaptation policies is linked to the way the public perceives (1) the level of consequences from possible environmental consequences and (2) the level of threat resulting from global climate change.
- The public's position towards climate change is the main reason for the (1) low policy support, (2) willingness to pay for climate change policies, and (3) willingness to change their behavior related to mitigation and adaptation.

Policy support and perceived levels of consequences from future environmental changes

Table 5.1 shows the results of the first cross tabulation that tests the relationships between degree of support for climate change mitigation and adaption policies and perceive level of consequences from environmental changes for all nine countries combined. On the global scale with the data of all nine countries combined, the chi-square test was found to be $p < 0.0005$. This suggests a statistical significant relationship between support for mitigation and adaptation strategies and perceived level of consequences from future environmental changes. For example, 49.2 percent of the people who strongly oppose mitigation policies and 44.6 percent who strongly oppose adaptation policies also believe that climate change and other environmental changes are either not happening at all or do not result in any negative consequences. Furthermore, 48.8 percent of the participants who strongly or moderately oppose mitigation policies and 57.9 percent who strongly or moderately oppose adaptation policies also expect slight negative consequences from environmental changes over the next 20 years. On the other end of the scale, 91.7 percent of the people who strongly support mitigation policies and 92.1 percent who strongly support adaptation polices also expect serious negative consequences. This means that if people believe that future changes to the environment have serious negative consequences they are more likely to support mitigation and adaption strategies. This global trend is supported by the data for the individual nine countries as well. Over 80 percent of the participants in Brazil, Canada, Germany, Japan, Mexico, the Netherlands, Spain, the United Kingdom, and the United States who strongly support mitigation or adaptation policies also expect serious negative consequences from environment changes.

For Canada, Germany, and the United Kingdom, the data indicate approximately 70 percent of the public moderately supporting mitigation or adaptation

Table 5.1 Global trends for the relationship between public support of climate change policies and the perceived level of consequences from environmental changes

		Index of Support for Mitigation Policies				
		Strongly Oppose	Moderately Oppose	Undecided	Moderately Support	Strongly Support
Public's perceived level of consequences from environmental changes	not likely at all to happen	23.5%	2.5%	0.8%	0.1%	0.5%
	no neg. cons.	25.9%	9.6%	2.1%	0.1%	0.4%
	slightly neg. cons.	23.5%	25.3%	13.8%	2.5%	1.0%
	moderate neg. cons.	18.5%	36.3%	28.5%	16.9%	6.5%
	serious neg. cons.	8.6%	26.4%	54.9%	80.4%	91.7%

		Index of Support for Adaptation Policies				
		Strongly Oppose	Moderately Oppose	Undecided	Moderately Support	Strongly Support
	not likely at all to happen	18.9%	1.4%	1.0%	0.1%	0.4%
	no neg. cons.	25.7%	9.5%	2.3%	0.2%	0.2%
	slightly neg. cons.	26.1%	31.8%	16.4%	2.6%	1.3%
	moderate neg. cons.	20.3%	35.5%	33.3%	19.5%	6.1%
	serious neg. cons.	9.0%	22.0%	47.0%	77.6%	92.1%

Note: Y-axis, survey participants were asked on a scale 1 from 5, where 1 is "not likely at all to happen" and 5 is "serious negative consequences", over the next 20 years what do you think the level of consequences will be if any, of a substantial increase in global warming resulting in climate change? X-axis, survey participants were asked on a scale from 1 to 5, where 1 is "strongly oppose" and 5 is "strongly support", how much do you support or oppose the following policies to reduce global climate change? The indexes were calculated based on the degree the participants supported the different mitigation and adaption policies.

policies also expects serious negative consequences from climate change and other environmental changes. In addition, in the cases of the public in Germany, the Netherlands, and the United States, the data show strong relationships between opposing any type of climate change policy and not expecting any environmental changes in the foreseeable future. For example, in Germany 46.2 percent, in the Netherlands, 15.4 percent, and in the United States 22.5 percent who strongly opposed mitigation policies also do not expect any environmental changes, including climate change, to occur within the next 20 years.

Furthermore, 42.9 percent of the respondents in Germany and 18.2 percent in the United States who strongly oppose adaption policies do not anticipate any environmental changes at all. Overall, the analysis identified several significant relationships for the nine surveyed countries. The data find strong correlations between the public's support for climate change mitigation and adaptation policies and their perceived level of consequences from environmental changes. People who strongly oppose climate change policies are also less likely to believe in negative consequences from environmental changes, whereas someone who is very supportive of mitigation and adaptation measures also tends to take changes to the environment very serious.

Policy support and perceived levels of threat resulting from climate change

Using the same methodology, a second crosstabulation was performed to test the relationship between mitigation/adaption policy support and perceived level of threat resulting from climate change and further test the underlying hypotheses:

- The public's general support for mitigation and adaptation policies is linked to (1) the way the public perceives the level of consequences from possible environmental consequences and (2) the level of threat resulting from global climate change.

Table 5.2 illustrates the relationships on the global scale using data from all nine countries combined into one sample. Similar to the contingency table discussed above, the chi-square test with $p < 0.0005$ suggests a statistical significant relationship between the two. The data show that about 46 percent of the population who strongly opposes mitigation and adaptation policies also perceives climate change as no threat at all. Also, over 60 percent who strongly or moderately oppose climate change policies also stated that they only view climate change as a slight threat. In terms of the people who are undecided in whether or not to support mitigation and adaptation polices, 43.2 percent perceive climate change as some threat.

Another significant relationship that was observed was between level of support for climate change policies and the belief that climate change poses a high level of threat. For this case, 83.2 percent of people who strongly support mitigation policies and 82.5 percent who strongly support adaptation polices perceive climate change as a high threat. On the international scale, five out of the nine countries show significant relationships between strong public support for mitigation policies and perceptions of climate

Table 5.2 Global trends for the relationship between public support of climate change policies and the perceived level of threat from climate change

		Index of Support for Mitigation Policies				
		Strongly Oppose	Moderately Oppose	Undecided	Moderately Support	Strongly Support
Public's perceived level of threat resulting from global climate change	no threat at all	46.3%	11.3%	2.5%	0.2%	0.1%
	a slight threat	30.2%	31.6%	13.5%	3.4%	0.7%
	some threat	13.0%	34.3%	43.2%	31.8%	15.9%
	a high threat	10.5%	22.8%	40.8%	64.6%	83.2%
		Index of Support for Adaptation Policies				
		Strongly Oppose	Moderately Oppose	Undecided	Moderately Support	Strongly Support
	no threat at all	46.4%	11.5%	2.1%	0.2%	0.2%
	a slight threat	31.1%	29.4%	17.2%	3.8%	1.0%
	some threat	13.1%	42.6%	46.7%	35.5%	16.3%
	a high threat	9.5%	16.6%	34.0%	60.5%	82.5%

Note: Y-axis, under the assumption that global climate change continues to occur over the next 50 years, survey participants were asked on a scale from 1 to 4, where 1 is "no threat at all" and 4 is "very high threat", how would you rate the level of threat resulting from global climate change? X-axis, survey participants were asked on a scale from 1 to 5, where 1 is "strongly oppose" and 5 is "strongly support", how much do you support or oppose the following policies to reduce global climate change? The indexes were calculated based on the degree the participants supported the different mitigation and adaption policies.

change as a high threat. For Canada, the data found that over 90 percent of the people who strongly support mitigation policies also perceive climate change as a high threat to plants and animals, people in other countries, people in their own country, or to themselves and their family. This is followed by the United States with 85.4 percent, Spain with 84.3 percent, and Germany with 82.6 percent. The fifth and last country where this relationship was statistical significant was the Netherlands where 45 percent of the participants strongly supporting mitigation policies perceived climate change as a high threat. As shown in Chapter 4, within these five countries, between 45 percent (Netherlands) and 65 percent (Spain) of the populations moderately or strongly support mitigation policies in general. Furthermore, 20 percent of the Spanish, 18 percent of the Canadian, 13 percent of United States, 12 percent of the Dutch, and 11 percent of the German population in general strongly supports mitigation policies.

All nine countries confirm significant relationships between strong support for the study's adaptation policies and perceiving climate change as a large threat. The data suggest that in Mexico 93.7 percent, in Brazil

91.1 percent, in the United States 86.6 percent, in Germany 85.9 percent, in Canada 84.7 percent, in Japan 80.8 percent, in Spain 79.6 percent, in the United Kingdom 75.3 percent, and in the Netherlands 38.1 percent of the public who strongly support adaptation policies also perceive climate change as a high threat. According to the frequency distributions between 48 percent (United States) and 86 percent (Mexico) of the public "moderately" or "strongly" support adaption policies in general. Moreover, over 40 percent in Brazil (45 percent) and Mexico (42 percent) show strong support for adaptation policies, compared to less than 30 percent in Canada (29 percent), Germany (25 percent), and Spain (23 percent). In the Netherlands, the United Kingdom, and the United States less than 20 percent of the public seems to strongly support adaptation policies.

Furthermore, the performed tests for statistical significance also showed that in Germany 61.5 percent of the participants who strongly opposed mitigation policies and 50 percent who strongly opposed adaptation policies also do not perceive climate change as a threat at all. The survey data from the Netherlands also show similar relationships. Within the Dutch population results indicate that 33.3 percent who strongly oppose mitigation policies and 40.6 percent who strongly oppose adaptation policies do not feel threatened by climate change. For the populations in Spain, the United Kingdom, and the United States, the data only show a relationship between strong opposition to adaptation policies and perceiving climate change as no threat at all. In Spain, 70 percent who strongly opposed adaptation policies also perceived climate change as no threat followed by the United States with 49.1 percent and the United Kingdom with 48 percent.

The data find strong positive correlations between the public's support for climate change mitigation and adaptation policies and their perceived overall level of threat from climate change. People who strongly oppose climate change policies are also very likely to perceive climate change as "no threat" or only "slight threat", whereas someone who is very supportive of mitigation and adaptation measures also tends to view climate change as a significant threat. This suggests as more people support mitigation and adaption policies more people will also consider climate change a significant threat. Overall, the analysis confirmed the hypotheses and identified several significant relationships for the nine surveyed countries between the public's general support for adaption and mitigation polices on the one hand and the perceived level of consequences from environmental changes and the level of threat resulting from climate change on the other.

Relationships between attitude, levels of concern, public support for climate policies, and willingness to commit to behavioral changes

The following discussion of the regression analyses is divided into different parts based on three groups of dependent variables, which were all tested with the same dependent variables presenting public's attitudes towards climate change and levels of concern. In particular, the second hypothesis, underlying the first research question listed at the beginning of this chapter, tested to what degree the public's lack of policy support, unwillingness to pay and commitment to behavioral changes can be explained by public attitudes and levels of concern towards global climate change. The public's preference towards four general climate change strategies and the level of belief in the reality of global climate change were used to describe a person's attitude towards climate change. The remaining independent variables focused on the public's level of concern regarding possible dangerous impacts of climate change on different geographical and personal levels as well as timescales.

General support for climate change policies

For six out of the nine countries, the results of the regression analysis show a strong and statistical significant relationship between the independent variables capturing the public's attitude and levels of concern towards climate change and the dependent variable presenting the public's support for mitigation policies in general. As already discussed in the previous chapter, the dependent variable was created from survey questions which asked the survey participants for their level of support for different mitigation policies on a 5-point scale ranging from "strongly oppose" to "strongly support". As illustrated in Table 5.3, the results show a strong relationship with R > 0.5 between the independent and dependent variable for the United States, Netherlands, Spain, United Kingdom, Germany, and Canada. Between 25.4 percent (Canada) and 44.7 percent (United States) of the variation in the public's overall support for mitigation can be explained by the independent variables. The stepwise regressions show that the level of concern regarding possible impacts of climate change is the strongest of the different independent or predictor variables for all six countries.

However, when looking at the second strongest dependent variable, the stepwise regressions identified country specific differences. In the cases of the Netherlands, Spain, and Germany, the second most influential independent variable is one of the attitudinal variables that asked the participants to choose between four general climate strategies. For the participants in the United Kingdom, the data show that the survey question asking how long it will take until climate change will be experienced somewhere on earth is the second strongest independent variable. For Canada the stepwise regressions demonstrate that the perceived level of threat of climate change over the next 50 years for oneself and family is the second strongest predictor variable for mitigation support, whereas in the United States the level of believe in the reality of climate change has the second strongest impact. For the three remaining countries Japan, Mexico, and Brazil, the R score is less than 0.5, indicating no large effect between the independent variables and the level of support for mitigation policies.

In terms of the relationship between the attitude and levels of concern towards climate change and support for adaptation policies the standard

Table 5.3 Strong relationships between (a) the independent variables capturing the public's attitude and levels of concern towards climate change and (b) the dependent variable presenting the public's support for mitigation policies in general

	Standard Regression				Stepwise Regression		
	Model		ANOVA		Strongest Variables	Model	
	R	Adj. R²	F	Sig.		R	Adj. R²
United States	.671	.447	128.6	.000	concern of pos. impacts	.626	.392
					& level of belief in CC★	.653	.426
Netherlands	.636	.401	97.5	.000	concern of pos. impacts	.566	.319
					& pref. general strategy	.613	.374
Spain	.596	.350	74.6	.000	concern of pos. impacts	.530	.280
					& pref. general strategy	.575	.329
United Kingdom	.570	.320	64.5	.000	concern of pos. impacts	.534	.284
					& impacts exp. on earth	.554	.305
Germany	.523	.268	51.3	.000	concern of pos. impacts	.478	.228
					& pref. general strategy	.508	.256
Canada	.512	.254	31.5	.000	concern of pos. impacts	.469	.219
					& concern for family	.490	.237

Note: CC stands for climate change.

regressions identified strong relationships for seven of the nine surveyed countries. The dependent variable, the index for the public's overall support for adaptation policies, is based on single survey questions, which asked the survey participants for their level of support for different specific adaptation policies on a 5-point scale ranging from "strongly oppose" to "strongly support". The frequency distribution and a more detailed discussion of the creation of this index were provided in the previous chapter. As shown in Table 5.4, the results of the country specific standard regressions confirmed significant and strong relationships with $R > 0.5$ between the predictor and outcome variables for the samples from the United States, the United Kingdom, Germany, Spain, the Netherlands, Canada, and Japan.

As a result, the variation in the public's overall support for adaptation policies can be explained to 57 percent in United States, to 45.1 percent in the United Kingdom, to 39.8 percent in Germany, to 39.5 percent in Spain, to 33.7 percent in the Netherlands, to 31.7 percent in Canada, and to 27.4 percent in Japan by the predictor variables. Again, the ensuing stepwise regression identified the level of concern variable as the predictor variable with the strongest relationship to the dependent variable, in this case the public's overall support for adaptation policies. Furthermore, for six of the seven countries who showed

Table 5.4 Strong relationships between (a) the independent variables capturing the public's attitude and levels of concern towards climate change and (b) the dependent variable presenting the public's support for adaptation policies in general

| | Standard Regression | | | | Stepwise Regression | | |
| | Model | | ANOVA | | Strongest Variables | Model | |
	R	Adj. R²	F	Sig.		R	Adj. R²
United States	.757	.570	209.9	.000	concern of pos. impacts	.706	.498
					& pref. general strategy	.731	.534
United Kingdom	.675	.451	111.7	.000	concern of pos. impacts	.616	.379
					& pref. general strategy	.646	.416
Germany	.634	.398	91.7	.000	concern of pos. impacts	.598	.357
					& pref. general strategy	.617	.379
Spain	.632	.395	90.2	.000	concern of pos. impacts	.568	.321
					& pref. general strategy	.611	.372
Netherlands	.619	.379	88.9	.000	concern of pos. impacts	.562	.315
					& pref. general strategy	.602	.361
Canada	.570	.317	42.7	.000	concern of pos. impacts	.519	.268
					& pref. general strategy	.547	.297
Japan	.524	.274	51.8	.000	concern of pos. impacts	.421	.176
					& concern for family	.470	.219

a strong relationship between the attitude and levels of concern towards climate change and support for adaptation policies, the attitudinal variable asking the participants to choose between four general climate strategies is the second strongest predictor variable. Only the Japanese sample identified the perceived level of threat of climate change over the next 50 years for oneself and family as the second strongest predictor variable for adaptation policy support.

Due to the identified strong relationships, the previous two overall indexes of mitigation and adaptation support were further broken into three more specific thematic indexes. This allowed testing the relationship between the independent variables and the public's support for energy efficiency policies, economic incentives, and for planning and adaptation strategies. All three sub-indexes are based on survey questions also used for the creation of the indexes of the public's support for mitigation and adaptation policies. The data found a strong relationship between the predictor variables and the sub-index of the public's support for energy efficiency policies for four of the nine countries. Based on the results of the standard regression analysis the answers provided by the participants show a strong relationship between the predictor and outcome variable in the United States (R = 0.627 and Adj. R^2 = 0.390), in the Netherlands (R = 593 and Adj. R^2 = 0.347), in Spain (R = 539 and Adj. R^2 = 0.286), and in Germany (R = 0.530 and Adj. R^2 = 0.276). Unfortunately, the standard regression does not show a strong relationship for the remaining five countries the United Kingdom, Canada, Japan, Mexico, and Brazil.

On the global scale, however, the regression analyses show a large effect of R = 0.545 and R^2 = 0.297 of the predictor variables on the outcome variable with the level of concern variable and the attitudinal variable capturing the public's preference of four different general climate change strategies having the strongest impact on the public's level of support for energy efficiency policies. In the cases of the United States, the Netherlands, Spain, and Germany, the stepwise regressions also identified the level of concern variable as the most influential variable. In addition, with the exception of the United States, the public's choice regarding the four general climate change policies is the second strongest independent variable. The data from the United States show that the level of belief in the reality of climate change is the second strongest variable after the public's level of concern regarding possible impacts of climate change.

Furthermore, the data from the United States and the Netherlands also indicate a strong relationship between the predictor variables and the second sub-index capturing the public's support for economic incentives. The data indicate that among the public in the United States 29.8 percent (R = 0.550 and Adj. R^2 = 0.298) and in the Netherlands 30.9 percent (R = 0.560 and Adj. R^2 = 0.309) of the variation in the public's support for economic incentives can be explained trough the predictor variables. In addition, at least the United Kingdom (R = 0.485 and Adj. R^2 = 0.229), Spain (R = 0.480 and Adj. R^2 = 0.229), Canada (R = 0.420 and Adj. R^2 = 0.167), and Germany (R = 0.385 and Adj. R^2 = 0.142) show a medium relationship between the attitude and levels of concern towards climate change and support for economic incentives. For both countries, United States and Netherlands, the performed stepwise regressions confirm the level of concern variable as the predictor variables with the strongest relationship to the dependent variable. The second strongest variable for the Netherlands is the public's attitude towards four general policies listed in Table 4.14 (Chapter 4) and for United States the level of belief regarding the reality of climate change shown in Table 4.13 (Chapter 4).

For the third sub-index, the public's support for planning and adaptation policies, the predictor variables showed a strong effect with R > 0.5 for the data collected from the United States, the United Kingdom, the Netherlands, Germany, Spain, and Canada. As displayed in Table 5.5, the predictor variables seem to have the strongest effect among the public in the United States, followed by the United Kingdom, Spain, the Netherlands, Germany, Canada, and Japan.

Thus, among the public in the United States 56.8 percent, in the United Kingdom 43.4 percent, in Spain 39.6 percent, in the Netherlands 39.5 percent, in Germany 35.7 percent, in Canada 31.4 percent, and in Japan 25.6 percent of the variation in the public's overall support for planning and adaptation strategies can be explained by the six independent variables addressing the attitude

Table 5.5 Strong relationships between (a) the independent variables capturing the public's attitude and levels of concern towards climate change and (b) the dependent variable presenting the sub-index for the public's support for planning and adaptation policies

	Standard Regression				Stepwise Regression		
	Model		ANOVA		Strongest Variables	Model	
	R	Adj. R²	F	Sig.		R	Adj. R²
United States	.756	.568	208.7	.000	concern of pos. impacts	.704	.495
					& concern for family	.729	.531
United Kingdom	.662	.434	104.1	.000	concern of pos. impacts	.613	.375
					& level of belief in CC★	.634	.400
Spain	.633	.396	90.5	.000	concern of pos. impacts	.562	.315
					& pref. general strategy	.615	.376
Netherlands	.632	.395	95.1	.000	concern of pos. impacts	.580	.335
					& pref. general strategy	.612	.373
Germany	.601	.357	77.2	.000	concern of pos. impacts	.567	.320
					& impacts exp. on Earth	.584	.340
Canada	.567	.314	42.0	.000	concern of pos. impacts	.515	.264
					& pref. general strategy	.537	.288
Japan	.511	.256	48.3	.000	concern of pos. impacts	.400	.159
					& concern for family	.459	.209

Note: CC stands for climate change.

and level of concern regarding possible negative impacts of climate change. The stepwise regression suggests that for all of these six countries the level of concern variable has the strongest relationship with the dependent variable. Ranging from $R = 0.704$ and Adj. $R^2 = 0.495$ in the case of the United States to $R = 0.400$ and Adj. $R^2 = 0.159$ for Japan.

Based on the second strongest independent variable, the seven countries can be organized into four different groups. The largest group consists of Spain, the Netherlands, and Canada for which the stepwise regressions identified the attitudinal variable asking the participants to choose between four general climate strategies as the second strongest independent variable. The second group includes the United States and Japan. In both cases, the variable capturing the participants level of concern for themselves and their family was the second strongest predictor variable. The third and fourth groups only consist of one country. The second strongest independent variable in the United Kingdom is the level of belief regarding the reality of climate change, whereas for Germany, the data show that the survey question asking how long it will take until climate change will be experienced somewhere on earth is the second strongest independent variable.

Similar to the precious indexes neither the regressions for Mexico nor Brazil shows a strong relationship (R > 0.5) between the independent variables and the outcome variable. This suggests that neither attitudes nor levels of concern seem to be major aspects during the public's decision process of supporting or opposing mitigation and adaptation policies.

Willingness to pay more for climate change abatement

The standard regression analyses between the predictor variables and the overall index of the public's willingness to pay more for climate strategies did not identify any large effects for any of the nine countries. As illustrated in Table 5.6, the conducted standard regressions only confirmed, at best, a medium relationship with R > 0.3 between the predictor and the outcome variables for the United States, the Netherlands, the United Kingdom, Canada, and Japan. Among these five countries. the public's attitudes and levels of concern can account for between 9.3 percent (Japan) and 18.4 percent (United States) of the variation in the public's willingness to pay more for climate strategies.

Thus, the data indicate that the independent variables do influence the public's willingness to pay more for climate change strategies, but are not the main or most important criteria the decision is based on. The public's preference towards four fundamental climate change strategies, the level of belief in the realty of climate change, the perception of how long it will take until dangerous

Table 5.6 Relationship between the public's attitude and levels of concern towards climate change and willingness to pay more for climate change strategies

| | Standard Regression | | | |
| | Model | | ANOVA | |
	R	Adj. R^2	F	Sig.
United States	.429	.184	35.318	.000
Netherlands	.398	.153	26.943	.000
United Kingdom	.393	.148	24.446	.000
Canada	.382	.137	15.191	.000
Japan	.300	.083	13.502	.000
Spain	.259	.060	9.770	.000
Germany	.248	.055	8.956	.000
Mexico	.190	.029	5.101	.000
Brazil	.112	.013	1.682	.122

impacts of climate change will be experienced on earth and in their region, the level of concern regarding the possible impacts of climate change, and the perceived level of threat resulting from climate change for themselves and their families does not largely influence the public's willingness to pay more for climate change mitigation or adaptation policies. Instead, other factors may explain the public's willingness to pay more for climate change policies which were not captured by the survey instrument. Subsequently, the regression analyses between the six predictor variables and all the sub-indexes such as the willingness to pay more for public transit, renewable energy, or taxes did also not show any large effects among the nine countries.

Willingness to change behavior

Another strong relationship was established between the predictor variables and the overall index for the public's willingness to change their behavior to reduce the causes and impacts of climate change. The outcome or dependent variable is based on the responses to single survey questions that asked the participants for their level of willingness to change their behavior in such areas as to use public transit for most of their travel, install solar panels on their home, buy mainly locally produced goods, use mainly recycled paper, purchase only energy saving appliances, and insulate their home or apartment. The global frequency distribution of this index suggests that close to 80 percent of all survey participants are in principle willing to strongly willing to change their behavior and thus live a more sustainable lifestyle.

The standard regression analysis indicates a strong relationship of R > 0.5 among the participants from the United States (R = 0.573 and Adj. R^2 = 0.324), the United Kingdom (R = 0.557 and Adj. R^2 = 0.305), Germany (R = 0.538 and Adj. R^2 = 0.284), and Spain (R = 0.524 and Adj. R^2 = 0.269). Among the public in the United States, 32.4 percent, in the United Kingdom 30.5 percent, in Germany 28 percent, and in Spain 26.9 percent of the variation in the public's willingness to change their behavior can be explained by the predictor variables. As shown in Table 5.7, the stepwise regressions once again suggest that for all four countries the level of concern variable has the strongest relationship with the dependent variable. This independent variable is followed by the attitudinal variable asking the participants to choose between four general climate strategies the in the cases of the United Kingdom, Germany, and Spain. For the United States, the variable capturing the participants level of concern for themselves and their family was the second strongest predictor variable.

The regression analyses for the remaining five countries Canada, Netherlands, Japan, Brazil, and Mexico showed an R score of less than 0.5 indicating no large effect between attitude and levels of concern towards climate change and the public's willingness to change their behavior. However, the regressions of the remaining five countries all show a R score of R > 0.3 indicating at the least a medium relationship between the public's attitudes and levels of concern

Table 5.7 Strong relationships between (a) the independent variables capturing the public's attitude and levels of concern towards climate change and (b) the dependent variable presenting the overall index for the public's willingness to change their behavior to reduce the causes and impacts of climate change

	Standard Regression				Stepwise Regression		
	Model		ANOVA		Strongest Variables	Model	
	R	Adj. R²	F	Sig.		R	Adj. R²
United States	.573	.324	76.447	.000	concern of pos. impacts	.547	.298
					& concern for family	.561	.314
United Kingdom	.557	.305	60.098	.000	concern of pos. impacts	.516	.265
					& pref. general strategy	.546	.297
Germany	.538	.284	55.481	.000	concern of pos. impacts	.516	.265
					& pref. general strategy	.530	.279
Spain	.524	.269	51.334	.000	concern of pos. impacts	.469	.219
					& pref. general strategy	.517	.266

towards climate change on the one hand and their willingness to commit to behavioral changes on the other. Despite only four individual countries showing strong relationships the standard regressions for the global scale with all nine countries combined show an R score of R > 0.5. This means that the global sample does show a strong relationship (R = 0.543 and Adj. R² = 0.295) between the independent variables and the public's willingness to change their behavior. Therefore, the data suggest that on the global scale, 29.5 percent of the variation in the dependent variable can be explained by the dependent variables. In particular, the answers provided to the two survey questions measuring the public's general level of concern and preferences regarding four fundamentally different climate change strategies seem to have the strongest impact on whether or not someone is willing to commit to behavioral changes.

The role of socioeconomic characteristics

To examine the role of socioeconomic variables the following two hypotheses were tested through regression analyses following the same methodology applied in the previous paragraphs of this chapter:

- The general attitude towards climate change is impacted by socioeconomic variables.
- Public risk perceptions of climate change are significantly impacted by socioeconomic variables.

Socioeconomic variables and general climate change attitudes

Can differences in general attitudes towards climate change among the nine countries be explained to a large degree by socioeconomic characteristics? As discussed earlier, the public's general attitude towards climate change was measured through level of believe in the reality of climate change and their preference regarding the following four fundamental climate change policies.

- "We should not take any steps that would have economic costs until we are certain that climate change is really a problem."
- "We should take some steps just in case climate change is real."
- "We only should take steps to address climate change which are low in costs."
- "Climate change is a serious problem, and we should begin taking steps now even if this involves significant cost."

Standard and stepwise regressions were conducted to determine the effect of independent socioeconomic variables such as age, gender, household income, and education on the two dependent variables presenting the public's attitude towards climate change.

The regression analysis did not indicate any large effects between the socioeconomic variables and the two dependent attitudinal variables. As shown in Table 5.8, the statistical tests of the standard regressions only, at best, showed weak significant relationships between the variables with $R > 0.2$ and $R^2 > 0.03$. That is, the results suggest that gender, household income, level of education, or age only account for a small percentage of the variation in the public's level of belief and preference towards general climate change strategies.

Socioeconomic variables and climate change risk perceptions

The second hypothesis is also not confirmed by the results of different regression analyses. Compared to the hypothesis discussed in the previous section, regression analyses were performed to examine to what degree the four socioeconomic variables (age, gender, level of education, and household income) can explain differences in the public's risk perception of climate change among the nine countries. The results show that socioeconomic variables are not a strong predictor of perceived risks of global climate change. Instead, the calculated R and R^2 scores showed only a small correlation between the independent and dependent variables. At no point, does the R score reach 0.3 indicating at least a medium relationship between the predictor and outcome variables. Instead, the data show that the socioeconomic characteristics do not have a significant impact on the way climate change risks are perceived.

Table 5.8 Regression results for relationships between socioeconomic characteristics and attitude towards climate change policies

Relationship Between Socioeconomic Characteristics and Preference Towards Four General Global Climate Change Strategies

	Model		ANOVA	
	R	Adj. R²	F	Sig.
Brazil	.188	.030	6.423	.000
United Kingdom	.187	.030	6.438	.000
Mexico	.159	.018	3.544	.007
United States	.157	.020	5.550	.000
Germany	.155	.018	4.145	.003
Japan	.145	.016	3.906	.004
Spain	.134	.012	2.937	.020
Netherlands	.104	.005	1.789	.129
Canada	.095	.002	1.216	.303

Relationship Between Socioeconomic Characteristics and Level of Belief in the Reality of Global Climate Change

	Model		ANOVA	
	R	Adj. R²	F	Sig.
United Kingdom	.228	.047	9.683	.000
United States	.199	.035	9.045	.000
Netherlands	.174	.024	5.101	.000
Spain	.148	.016	3.628	.006
Japan	.145	.016	3.908	.004
Mexico	.143	.013	2.853	.023
Brazil	.123	.010	2.701	.030
Germany	.109	.006	2.030	.089
Canada	.103	.003	1.421	.226

Trust factors and public climate change risk perceptions

Studies show that public distrust in individuals, industries, governmental departments, and other institutions of organizations involved in risk and hazard management is strongly linked to risk perceptions (Bord & O'Connor, 1990; Mushkatel & Pijawka, 1992; Flynn et al., 1995). Typically, the more the public distrusts risk management and communicators information, the more concern they have about adverse impacts and potential threats for their own well-being (Slovic et al, 1991).

Research also shows that the failures in risk communication is significantly influenced by the public's trust in the communicator and in the ability of individuals, industries, or institutions responsible for risk management (Renn & Levine, 1991; Kasperson et al., 1992; Nye Jr. et al., 1997). Where risks are characterized by high uncertainties, as in climate change, trust may play a critical role in the success of risk communication and implementation of policies. Moreover, trust is not only a necessary precondition for successful climate change communication, but it can also be improved by well-developed communication strategies (Misztal, 1996). Trust is vital in organizations whose risk management policies impact communities and the environment in order to reduce complexity and generate social co-operation (Cvetkovich & Loefstedt, 1999).

However, much of this research was not conducted for hazards with high levels of uncertainties such as climate change, but in the context of technological risks such as nuclear power. Being important, this led to the question if such relationship between trust factors and public risk perceptions also exist in the context of climate change. Therefore, two different groups of regression analyses were conducted. The first group of regressions focused on the relationship between public trust in the science of climate change, as well as towards different sources of climate change information, and the level of concern over the impacts of climate change. This relationship is of great importance for communication efforts in order to improve public policy decision making. As shown in the first half of this chapter, the public's level of concern over possible climate change impacts has the strongest effect on the level of support for mitigation and adaptation policies among all tested climate change risk perceptions. Moreover, as discussed earlier, success of commutation efforts is significantly impacted by the public's trust in the communicator such as family and friends, mainstream media, governmental and environmental organizations, scientists and teachers, and corporations. The results of the regressions are discussed in the next section.

The second group of regressions focused on the effect of climate change threat and risk perceptions on trust perceptions in the federal government as a potential communicator and risk manager. Due to the government's capability to implement the needed, comprehensive climate change policies, trust in the government is of significant importance to successfully mitigate and adapt to climate change. Furthermore, the survey data show that the public perceives the national or federal government as the primary institution responsible to reduce or mitigate climate change. The public's perceived level of responsibility of the federal government to reduce climate change was used as a surrogate variable for the public's level of trust in the climate change risk management capabilities of the government. This is based on the rationale that the public would not perceive the government as responsible to reduce climate change without acknowledging its capability to do so in the first place.

In addition, two additive indexes were created to present the public's perceived risks: (1) general level of threat from climate change, and (2) risk of climate change causing negative impacts over the next 50 years. The additive index of the public's perceived level of threat resulting from climate change is based on how the survey participants rated the threat level of climate change for plants and animals, people in other countries, people in their country, and for themselves and their family over the next 50 years. The additive risk index is based on how the respondents rated the risk of climate change causing different environmental and societal impacts over the next 50 years.

Trust factors and level of concern over climate change impacts

As mentioned above, regressions were used to test the relationships between the public's trust in the science of climate change, as well as towards different sources of climate change information, and the level of concern over the possible impacts of climate change. Table 5.9 summarizes the results of the standard regressions on the global scale with all nine countries combined into one sample. The results show a strong relationship (R = 0.569 and Adj. R^2 = 0.322) between the independent trust variables and the public's level of concern over climate change. The data show that 32.2 percent of the variation in the public's

Table 5.9 Standard regression results for the relationship between (a) trust in science as well as towards different sources of information and (b) level of concern for all countries combined

Model				Coefficients				
	ANOVA			Independent Variable	Unstand.	Stand.	t	Sig.
R	Adj.R²	F	Sig.		B	Beta		
.569	.322	314.913	.000	enough scientific data	−.051	−.050	−4.380	.000
				scientific data are trustworthy	.276	.245	18.215	.000
				trust TV weather reports	.028	.024	1.954	.000
				trust corporation	−.017	−.015	−1.171	.241
				trust family and friends	.102	.080	7.588	.000
				trust governmental organizations	−.113	−.106	−7.887	.000
				trust environmental organizations	.345	.342	23.465	.000
				trust mainstream news media	−.013	−.012	−.840	.401
				trust scientists	.134	.113	8.002	.000
				trust religious leaders	−.016	−.016	−1.391	.164
				trust teachers	.032	.027	2.089	.037

level of concern can be explained by their level of trust towards climate change science and different sources of information. The results also show that the relationship between the different independent variables and the one dependent variable are not always positive. For example, the data indicate that the more people consider the climate change scientific data as trustworthy, or trust environmental organizations the more they are concerned about climate change. On the other hand, the more the public trusts the information from corporations or governmental organizations, the less concerned are they about climate change.

As already discussed in Chapter 4, the majority of the public is concerned about global climate change and its potential impacts; 83.7 percent stated at least some concern regarding the possible impacts of climate change. Regarding the independent trust variables, corporations, governmental organizations, and religious leaders are trusted the least as a source for climate change information. Instead, the majority of the public somewhat or strongly trusts scientists (66.8 percent), television weather reports (61.3 percent), family and friends (56.5 percent), and environmental organizations (56.1 percent) as sources of climate change information. The results also show that a large numbers of people are undecided in whether or not they should trust certain sources of information. Especially, in regards to teachers, 40 percent of all participants chose the answer category "undecided" followed by corporations (36 percent), family and friends (33.1 percent), and the mainstream news media (30.5 percent). The frequency distributions among the nine countries also indicate an inconsistency between the level of trust towards of scientists and the level of agreement with the sufficiency and trustworthiness of scientific data and expert knowledge: 66.8 percent of the total sample "somewhat" or "strongly" trusts scientists as sources for climate change information. However, the data also indicate that 40.2 percent of the total sample doubts that the scientific community actually has enough data to fully understand climate change.

On the international scale, the regression analysis shows a strong and statistical significant relationship between the independent variables capturing the public's trust towards climate change science as well as potential communicators and the level of concern about possible impacts of climate change. Table 5.10 shows a strong relationship of $R > 0.5$ between the independent and dependent variables for the United States, United Kingdom, Netherlands, and Canada. According to the adjusted R^2 scores between 28.8 percent (Canada) and 54 percent (United States) of the variation in the public's level of concern can be explained by the independent variables.

According to the stepwise regressions the two strongest independent variables are the level of trust in environmental organizations and perceived trustworthiness of climate change science. Both variables have a positive relationship to the dependent variable. Thus, the higher the trust in environmental organizations and the scientific data the more concerned is the public about the possible impacts of climate change.

Table 5.10 Countries with strong relationships between (a) trust towards climate change science as well as communicators and (b) level of concern about possible impacts of climate change

	Standard Regression				Stepwise Regression		
	Model		ANOVA		Strongest Variables	Model	
	R	Adj. R²	F	Sig.		R	R²
United States	.738	.540	101.9	.000	trustworth. of CC science	.671	.339
					& trust environ. org.	.723	.522
United Kingdom	.640	.402	50.3	.000	trust environ. org	.578	.334
					& trustworth. of CC science	.613	.374
Netherlands	.590	.340	41.4	.000	trust environ. org.	.547	.298
					& trustworth. of CC science	.574	.328
Canada	.550	.288	20.8	.000	trust environ. Org.	.434	.187
					& trustworth. of CC science	.486	.233

Note: CC stands for climate change.

Impact of climate change risk and threat perceptions on trust in government as source of information and risk manager

The second group of regressions focused on the impact of risk and threat perceptions on the public's trust in the federal government as source of climate change information and as risk manager capable of implementing successful mitigation and adaptation strategies. On the global scale, the regressions show only a weak relationship ($R < 0.3$) between the public's perceived level of threat from climate change as well as risks of climate change causing negative impacts over the next 50 years and the level of trust in government as a source of climate change information. This shows that how the public perceives the threat of climate change and the risks of potential negative impacts does not have a strong influence on the level of trust towards the government as a source of climate change information and potential speaker of communication programs. Regressions on the national scale for each of the nine countries also show no strong relationships of $R > 0.5$.

As shown in Table 5.11, the regression testing the strength of the relationship between the two risk and threat indexes and the public's trust in the government as risk manager also did not show a strong relationship ($R < 0.5$). However with an R score of $R = 0.427$ the regression shows a moderate relationship on the global scale. Furthermore, the analysis shows a positive relationship for both indexes and the public's level of trust in the government's capability as

Table 5.11 Impacts of climate change risk and threat perceptions on the public's level of trust in the government as a CC risk manager

Model			ANOVA		Coefficients				
R	R²	Adj.R²	F	Sig.	Ind. Var.	Unstand.	Stand.	t	Sig.
						B	Beta		
.427	.183	.182	810.319	.000	CC threat index	.289	.232	15.731	.000
					CC risk index	.255	..232	15.755	.000

Note: CC stands for climate change.

climate change risk manager. This indicates that the stronger the public believes in climate change risks and threats the more they trust the government to be capable of implementing successful mitigation and adaption policies.

On the international scale, the regression analysis shows a strong relationship only for the data from the United States. With an R score of R = 0.553 and a R² score 0.306. The result shows that the two indexes explain 30.6 percent of the public's variation of trust towards the government as a climate change risk manager. For Japan and Germany, the regressions showed moderate relationships as well. For the remaining countries, the R is below 0.3 indicating only a weak relationship between climate change risk and threat perceptions and trust in the climate change risk management capability of the government.

Impact of knowledge, trust, and responsibility factors on public support for climate change policies and strategies

This section focuses on the final research question of the "Global Survey on Public Attitudes towards Climate Change" research project:

• What role do level of knowledge and perceptions of trust and responsibility play in the public's level of support towards adaptation and mitigation policies?

The following paragraphs address the regression analysis focusing on the impact of level of knowledge, level of trust towards sources of information, and perceived levels of responsibilities of different groups for reducing the main causes of climate change on the public's policy support for mitigation and adaption. In particular, the different regressions analyzed the impact of the independent variables on the public's level of support for adaptation and mitigation policies, willingness to pay more for climate change abatement, and willingness to commit to different behavioral changes.

Support for mitigation and adaptation policies

The results of the regression analysis show strong and statistical significant relationships between the independent variables – knowledge, trust, and responsibility – and the dependent variable, the public's support for mitigation policies in general (for at least seven out of the nine countries). As illustrated in Table 5.12, the results show a strong relationship (with R > 0.5) between the independent and dependent variables for the United States, Netherlands, Spain, United Kingdom, Canada, Japan, and Germany. Between 28.4 percent (Germany) and 54.2 percent (United States) of the variation in the public's overall support for mitigation can be explained by the independent variables.

The stepwise regressions identified two independent variables as the strongest: the level of trust towards environmental organizations as sources of information and the perceived level of personal responsibility to reduce climate change, especially in Japan and Germany. For Mexico and Brazil, the R score is less than 0.5 indicating there is no large effect between the independent variables and the level of support for mitigation policies.

In terms of the relationship between the same independent variables and adaptation policies, the standard regressions identified strong relationships for the same seven countries as in the previous regression. Thus, the results show a strong relationship for the United States (R = 0.799 and Adj. R^2 = 0.542), United Kingdom (R = 0.653 and Adj. R^2 = 0.449), Spain (R = 0.661 and Adj. R^2 = 0.419), Germany (R = 0.651 and Adj. R^2 = 0.405), Netherlands (R = 0.643 and Adj. R^2 = 0.396), Canada (R = 0.611 and Adj. R^2 = 0.344), and Japan (R = 0.594 and Adj. R^2 = 333). Between 54.2 percent (United States) and 33.3 percent (Japan) of the variation in the public's overall support

Table 5.12 Significant relationships between (a) the independent variables knowledge, trust, and responsibility and (b) the dependent variable, the public's support for mitigation policies in general

	Standard Regression				Stepwise Regression		
	Model		ANOVA		Strongest Variable	Model	
	R	Adj. R^2	F	Sig.		R	R^2
United States	.736	.542	43.5	.000	trust environ. org.	.626	.391
Netherlands	.653	.426	24.9	.000	trust environ. org.	.549	.301
Spain	.646	.417	22.8	.000	trust environ. org.	.471	.221
United Kingdom	.601	.341	17.7	.000	trust environ. org.	.482	.231
Canada	.599	.359	12.0	.000	personal resp.	.449	.200
Japan	.559	.312	14.6	.000	personal resp.	.392	.152

for adaptation policies can be explained by the predictor variables. The ensuing stepwise regression identified differences among the countries in terms of the independent variable with the strongest effect on the public's level of support for adaptation policies in general. Again, in the cases of the United States, United Kingdom, and Netherlands the level of trust towards environmental organizations is the predictor variable with the strongest relationship to the dependent variable, in this case the public's overall support for adaptation policies. However, for the remaining four countries Spain, Germany, Canada, and Japan, the perceived levels of responsibility of the United Nations (Spain), Environmental Protection Agency (Canada), local government (Japan), and personally (Germany) are the strongest independent variables for adaptation policy support.

The previous two overall indexes of mitigation and adaptation support were further broken into three more specific thematic indexes: public's support for energy efficiency policies, economic incentives, and for planning and adaptation strategies. The regression analyses identified strong relationships for all three thematic indexes among the nine countries.

For the first sub-index, the public's support for energy efficiency policies, the regression analyses show a strong relationship between the predictor variables and the dependent variables for eight of the nine countries. Only the R score for Brazil is below 0.5 indicating no large effect between the independent variables and the level of support for energy efficiency policies. As shown in Table 5.13, for the remaining eight countries, the R scores range from 0.514 (Mexico) to 0.681 (United States). The adjusted R^2 scores suggest that the independent variables account for between 24.1 percent and 45 percent of the variation in the dependent variable. The stepwise regressions identified the level of trust towards environmental organizations for the United States and Netherlands and the level of trust towards family and friends for Spain as the most influential variables. In the cases of Germany, Canada, Japan, United Kingdom, and Mexico the perceived level of responsibility of corporations, personally, the Environmental Protection Agency, or the federal government have the strongest impact on the public's level of support for energy efficiency policies.

In the case of the second sub-index, the public support for economic incentives, the regression analyses show strong relationships between the predictor variables and the dependent variables for five of the nine countries. The survey included economic incentives such as the government requiring higher utility rates from using non-renewable energy sources or higher taxes on electricity. The data indicate that among the public in the United States, 42.8 percent (R = 0.555 and Adj. R^2 = 0.428), in the Netherlands 34.8 percent (R = 0.605 and Adj. R^2 = 0.348), in the United Kingdom 31.6 percent (R = 0.581 and Adj. R^2 = 0.316), in Spain 31.5 percent (R = 0.579 & Adj. R^2 = 0.315), and in Canada 23.4 percent (R = 0.518 and Adj. R^2 = 0.234) of the variation in the public support for economic

Table 5.13 Significant relationships between (a) the variables knowledge, trust, and responsibility and (b) the support for energy efficiency policies

	Standard Regression				Stepwise Regression		
	Model		ANOVA		Strongest Variable	Model	
	R	Adj. R²	F	Sig.		R	R²
United States	.681	.450	31.9	.000	trust environ. org.	.548	.299
Spain	.611	.353	18.9	.000	trust family and friends	.420	.175
Netherlands	.597	.338	18.7	.000	trust environ. org.	.470	.220
Germany	.581	.317	16.3	.000	corporate resp.	.383	.146
Canada	.580	.305	10.9	.000	personal resp.	.414	.170
Japan	.563	.296	14.9	.000	EPA* resp.	.405	.163
United Kingdom	.517	.244	11.5	.000	fed. govern. resp.	.362	.130
Mexico	.514	.241	11.5	.000	personal resp.	.325	.104

Note: EPA stands for Environmental Protection Agency.

incentives can be explained through the independent variables knowledge, trust, and responsibility. According to the stepwise regressions aspects such as the perceived trustworthiness of the scientific climate change data, trust in environmental organizations, as well as perceived corporate and personal responsibility are the strongest independent variables among the five countries.

For the third sub-index, the public's support for planning and adaptation policies, the predictor variables showed a strong effect with $R > 0.5$ for the data collected from the United States, Spain, United Kingdom, Netherlands, Canada, Germany, Japan, and Mexico. Again, Brazil is the only country with an R score below 0.5.

As displayed in Table 5.14, the independent variables combined seem to have the strongest effect among the public in the United States ($R = 0.804$ and Adj. $R^2 = 0.637$), followed by Spain ($R = 0.677$ and Adj. $R^2 = 0.441$),United Kingdom ($R = 0.677$ and Adj. $R^2 = 0.441$), Netherlands ($R = 0.649$ and Adj. $R^2 = 0.404$), Canada ($R = 0.625$ and Adj. $R^2= 0.362$), Germany ($R = 0.610$ and $R^2 = 0.352$), Japan ($R = 0.600$ and Adj. $R^2 = 0.340$), and Mexico ($R = 0.506$ and $R^2 = 0.232$). Thus, between 23.2 percent (Mexico) and 63.7 percent (United States) of the variation in the public's overall support for planning and adaptation strategies can be explained by the public's level of knowledge, level of trust towards climate change information as well as potential sources, and perceived levels of responsibilities of different groups for engaging in mitigative actions.

Table 5.14 Significant relationships between (a) the variables knowledge, trust, and responsibility and (b) the support for planning and adaptation policies

	Standard Regression				Stepwise Regression		
	Model		ANOVA		Strongest Variable	Model	
	R	Adj. R²	F	Sig.		R	R²
United States	.804	.637	67.5	.000	trust environ. org.	.695	.482
Spain	.677	.442	27.0	.000	UN resp.	.467	.217
United Kingdom	.677	.441	26.5	.000	trust environ. org.	.519	.269
Netherlands	.649	.404	24.4	.000	trust environ. org	.505	.255
Canada	.625	.362	13.7	.000	EPA★ resp.	.433	.186
Germany	.610	.352	18.9	.000	personal resp.	.431	.185
Japan	.600	.340	18.1	.000	loc. govern. resp.	.420	.175
Mexico	.506	.232	11.0	.000	personal resp.	.317	.099

Note: EPA stands for Environmental Protection Agency.

Willingness to pay more for climate change abatement

The standard regression analyses between the predictor variables and the overall index of the public's willingness to pay more for climate change abatement did only show a large relationship for the data collected form United States with $R = 0.507$ and Adj. $R^2 = 0.232$. As shown in Table 5.15, the regressions only confirmed, at best, a medium relationship with $R > 0.3$ between the predictor and the outcome variables for Canada, United Kingdom, Netherlands, Japan, Germany, Spain, and Brazil.

In the case of the United States, the independent variables can account for 23.2 percent of the variation in the public's willingness to pay more for climate strategies in general. Furthermore, the different regression analyses also show that the perceived trustworthiness of the scientific climate change data is the most influential independent variable for the United States. With the exception of the United States, the independent variables presenting the public's level of knowledge, level of trust towards sources of information, and perceived levels of responsibilities of different groups for reducing the main causes of climate change do not largely influence the public's willingness to pay more for climate change mitigation or adaptation policies.

Willingness to change behavior

Another strong relationship was established between the predictor variables and the overall index for the public's willingness to change their behavior to reduce the causes and impacts of climate change. As already discussed, the

Table 5.15 Relationship between (a) perceived levels of knowledge, trust, as well as responsibility and (b) the public's willingness to pay more for climate abatement in general

| | Standard Regression | | | |
| | Model | | ANOVA | |
	R	Adj.R²	F	Sig.
United States	.507	.232	12.777	.000
Canada	.499	.214	7.111	.000
United Kingdom	.488	.214	9.795	.000
Netherlands	.476	.204	9.865	.000
Japan	.400	.134	6.111	.000
Germany	.394	.128	5.851	.000
Spain	.367	.107	4.950	.000
Brazil	.302	.062	3.111	.000
Mexico	.266	.042	2.444	.000

outcome or dependent variable is based on the responses to six single survey questions that asked the participants for their level of willingness to change their behavior in such areas as to use public transit for most of their travel, install solar panels on their home, buy mainly locally produced goods, use mainly recycled paper, purchase only energy saving appliances, and insulate their home or apartment. The global frequency distribution of this index suggests that close to 80 percent of all survey participants are in principle willing to strongly willing to change their behavior and thus live a more sustainable lifestyle.

The standard regression analysis indicates a strong relationship of $R > 0.5$ among the participants from the United States ($R = 0.619$ and Adj. $R^2 = 0.366$), Spain ($R = 0.581$ and Adj. $R^2 = 0.317$), United Kingdom ($R = 0.577$ and Adj. $R^2 = 0.311$), Germany ($R = 0.572$ and Adj. $R^2 = 0.306$), Canada ($R = 0.536$ and Adj. $R^2 = 0.254$), Netherlands ($R = 0.506$ and Adj. $R^2 = 0.234$), and Japan ($R = 0.501$ and Adj. $R^2 = 0.228$). Thus, among these seven countries with $R > 0.5$, between 22.8 percent (Japan) and 36.6 percent (United States) of the variation in the public's willingness to change their behavior can be explained by the independent variables.

As shown in Table 5.16, the stepwise regressions shows level of perceived personal responsibility has the strongest relationship with the dependent variable in the cases of the United States, United Kingdom, Germany, and the Netherlands. For Spain, the data suggest the perceived level of community responsibility, and for Japan, corporate responsibility as the independent variables with the strongest impact on the public's overall willingness to commit to behavioral changes. In the case of Canada, however, the level of trust towards scientists as

Table 5.16 Significant relationships between (a) the variables knowledge, trust, and responsibility and (b) overall willingness to commit to behavioral changes

	Standard Regression				Stepwise Regression		
	Model		ANOVA		Strongest Variable	Model	
	R	Adj. R^2	F	Sig.		R	R^2
United States	.619	.366	22.9	.000	personal resp.	.491	.240
Spain	.581	.317	16.2	.000	community resp.	.380	.143
United Kingdom	.577	.311	15.6	.000	personal resp.	.434	.188
Germany	.572	.306	15.5	.000	personal resp.	.368	.135
Canada	.536	.254	8.6	.000	trust scientists	.333	.109
Netherlands	.506	.234	11.6	.000	personal resp.	.367	.134
Japan	.501	.228	10.8	.000	corporate resp.	.349	.120

sources of information is the most influential singe independent variable. The regression analyses for the remaining two countries, Brazil and Mexico, showed an R score of less than 0.5 indicating no large effect between the 25 independent variables and the public's willingness to change their behavior. However, the regressions show an R score of R > 0.3 indicating at the least a medium relationship.

The ensuing regression analyses between the independent variables and the single survey questions comprising the overall index show strong relationships for some of the individual survey questions and countries. Altogether, the index was created based on the responses to six single survey questions which asked the participants for their level of willingness to change their behavior in such areas as to use public transit for most of their travel, install solar panels on their home, buy mainly locally produced goods, use mainly recycled paper, purchase only energy saving appliances, and insulate their home or apartment. The regression analysis show strong relationships between the independent variables and the willingness to install solar panels on their home, use mainly locally produced goods, and to purchase only energy saving appliances.

Table 5.17 summarizes the significant result and displays the countries for which the regression analysis established a strong relationship between the independent variables and the particular survey questions function as dependent variables. As the table shows the regression analysis only established strong relationships between the independent variables and three out of the six individual survey questions. In case of the United States, the data show a strong relationships between the predictor variables and all three individual survey questions with R scores ranging from R = 0.522 (willingness to install solar panels on home) to R = 0.581 (willingness to use mainly locally produced goods). Data

Table 5.17 Strong relationships between (a) perceived levels of knowledge, trust, as well as responsibility and (b) specific behavioral survey questions

Willingness to Install Solar Panels on Home

	Standard Regression				Stepwise Regression		
	Model		ANOVA		Strongest Variable	Model	
	R	Adj. R²	F	Sig.		R	R²
United States	.522	.252	13.766	.000	personal resp.	.393	.153

Willingness to Use Mainly Locally Produced Goods

	Standard Regression				Stepwise Regression		
	Model		ANOVA		Strongest Variable	Model	
	R	Adj. R²	F	Sig.		R	R²
United States	.581	.320	18.783	.000	trust scientists	.441	.194
Spain	.547	.277	13.573	.000	trust scientists	.349	.121
Canada	.511	.227	7.582	.000	trust scientists	.327	.105

Willingness to Purchase Only Energy Saving Appliances

	Standard Regression				Stepwise Regression		
	Model		ANOVA		Strongest Variable	Model	
	R	Adj. R²	F	Sig.		R	R²
United States	.560	.295	16.804	.000	state govern. resp.	.425	.179
Germany	.553	.284	14.088	.000	trust scientists	.352	.123
United Kingdom	.522	.249	11.709	.000	personal resp.	.366	.133
Spain	.516	.244	11.569	.000	trust scientists	.340	.114

from Spain confirm a strong relationships between the independent variables capturing the public's level of knowledge, level of trust towards climate change information as well as potential sources, and perceived levels of responsibilities of different groups for engaging in mitigative actions and the two answers provided to the two survey questions assessing the public's willingness to use mainly locally produced goods (R = 0.547 and Adj. R² = 0.277) as well as to purchase only energy saving appliances (R = 0.516 and Adj. R² = 0.244).

For Germany and the United Kingdom, the regressions only identified a strong relationship between the independent variables and the survey question addressing the public's willingness to purchase only energy saving appliances. In particular, the analysis shows a relationship for Germany with R = 0.553 and for the United Kingdom with R = 0.522. Canada's data confirmed a relationship between the predictor variables and the answers provide to the survey question focusing on the public's willingness to use mainly locally produced goods with R = 0.511.

Overall, the calculated R^2 scores suggest that the independent variables can account between 22.7 percent and 32 percent of the variation of the answers provided to the three survey questions. Moreover, the stepwise regressions identify the level of trust in scientists as source of climate change information as the independent variable having most often the strongest impact on the dependent variables, followed by perceived personal responsibility to reduce climate change and perceived responsibility for the state government to engage in mitigative actions.

Summary

Several relationships between survey-based independent and dependent variables were explored in this chapter. People who already have experienced climate change or believe they will experience climate change soon are more concerned about it compared to people who believe they will not experience impacts from climate change in the future. In terms of the role climate change risk perceptions and attitudes play in the public's willingness to support mitigation and adaptation policies, the analysis identified several significant relationships for the nine surveyed countries. The data find strong correlations between the public's support for climate change mitigation and adaptation policies and their perceived level of consequences from environmental change as well as perceived level of threat from climate change. The data received from multiple countries also indicate that if people strongly oppose climate change policies, they also do not expect environmental changes such as climate change, worsening of urban air pollution, or an increasing frequency of major hurricanes and/ or floods to occur over the next 20 years, and generally perceive climate change as no threat at all.

Numerous multiple regressions were conducted focusing on the relationship between attitudes and levels of concern over climate change and willingness to support climate change mitigation and adaptation policies, to pay more for climate abatement, and to commit to behavioral changes such as using public transit for most of their travel, installing solar panels on their home, buying mainly locally produced goods, using mainly recycled paper, purchasing only energy saving appliances, and insulating their home and apartment. With the exception of Japan, Mexico, and Brazil the results of the regression analysis show a strong and statistical significant relationship (R > 0.5) between the independent variables capturing the public's attitude and levels of concern towards climate change and the dependent variable presenting the public's support for mitigation policies in general. In terms of the relationship between the attitude and levels of concern towards climate change and support for adaptation policies the standard regressions identified strong relationships for seven of the nine surveyed countries. Only for the samples of Brazil and Mexico was the R value below 0.5, indicating no strong correlation between the dependent and independent variables. This suggests that only in the cases of Brazil and Mexico neither attitudes nor levels of concern seem to be major aspects during the public's decision process of supporting or opposing mitigation and adaptation policies. The standard regression analyses between the predictor variables and the overall index of the public's willingness to pay more for climate abatement did not identify any large effects for any of the nine countries. A strong relationship was established between the predictor variables and the overall index for the public's willingness to change their behavior to reduce the causes and impacts of climate change. In particular the standard regression analysis indicates a strong relationship of $R > 0.5$ among the participants from the United States, the United Kingdom, and Spain. In addition for all of the remaining five countries Canada, the Netherlands, Japan, Brazil, and Mexico showed an R score of $R > 0.3$ indicating at the least a medium relationship between the public's attitudes as well as levels of concern towards climate change and their willingness to commit to behavioral changes.

Furthermore, the ensuing stepwise regressions identified the level of concern regarding possible impacts of climate change as the strongest of the independent or predictor variables for all countries and regressions. However, when looking at the second strongest dependent variable, the stepwise regressions identified country specific differences. The data identify, in most cases, the one of the attitudinal variables which asked the participant's to choose between four general strategies as the second most influential independent variable. Other less common independent variables which were identified by the stepwise regressions as second strongest predictor variables were the perceived level of threat of climate change over the next 50 years for oneself and family, the level of believe in the reality of climate change, and the survey question asking how long it will take until climate change will be experienced somewhere on Earth.

The data also show that characteristics such as age gender, household income, or education are do not influence someone's attitude or risk perception significantly towards climate change in any of the nine countries surveyed. Nevertheless, approximately one-third or more of the surveyed population believes that climate change poses a high risk for causing numerous negative environmental impacts. Furthermore, the data suggest that most people in the surveys are generally willing to support climate change mitigation through behavioral changes.

Regressions demonstrate a strong relationship between the independent trust variables and the public's level of concern over climate change. Roughly one-third of the variation in the public's level of concern can be explained by their level of trust towards climate change science and different sources of information. However, the regressions failed to show a strong relationship, on the global scale, between risk and threat perceptions on the public's trust in the federal government as source of climate change information and as risk manager. On the international scale the data only show a strong relationship for the United States between the two risk and threat indexes and the public's trust in the government as risk manager.

Such factors as knowledge, trust, and responsibility show several strong rela-tionships with climate change policy support for the nine countries. Strong relationship are seen between the independent variables and the public's support for mitigation and adaptation policies as well as their willingness to commit to behavioral change for seven out of the nine countries with Mexico and Brazil being the exception. The ensuing regression analyses between the independent variables and the single survey questions comprising the overall index for the public's willingness to change their behavior show strong relationships for some of the individual survey questions and countries. The regression analysis showed only strong relationships related to the willingness to install solar panels on their home, use mainly locally produced goods, and to purchase only energy saving appliances. Moreover, with the exception of the United States, the independ-ent variables do not largely influence the public's willingness to pay more for climate change mitigation or adaptation policies. The stepwise regressions did not identify any one single independent variable as the strongest for all seven countries. Instead, the independent variable with the strongest impact on the dependent variable varies by country. However, the variables presenting the perceived level of personal responsibility to reduce climate change and level of trust towards environmental organizations as a source of information seem to be the most influential.

Bibliography

Bord, R.J., & O'Connor, R.E. (1990). Risk communication, knowledge, and attitudes: Explaining reactions to a technology perceived as risky. *Risk Analysis, 10*, 499–506.

Cvetkovich, G., & Loefstedt, R. (Eds.). (1999). *Social Trust and the Management of Risk*. Lon-don, UK: Earthscan.

Field, A. (2009). *Discovering Statistics Using SPSS*. London, UK: Sage.

Flynn, J., Chalmers, J., Easterling, D., Kasperson, R.E., Kunreuther, H., Mertz, C.K., ... Slovic, P. (1995). *One Hundred Centuries of Solitude: Redirecting America's High-level Nuclear Waste Policies*. Boulder, CO: Westview Press.

Kasperson, R., Golding, D., & Tuler, S. (1992). Social distrust as a factor in sitting hazardous facilities and communicating risks. *Journal of Social Issues, 48*(4), 161–187.

Misztal, B.A. (1996). *Trust in Modern Societies*. Cambridge, UK: Polity Press.

Mushkatel, A.H., & Pijawka, K.D. (1992). *Institutional Trust, Information and Risk Perceptions: Report of Findings of the Las Vegas Metropolitan Area Survey*. Carson City: Nevada Nuclear Waste Project Office.

Nye Jr., J.S., Zelikow, P.D., & King, D.C. (1997). *Why People Don't Trust Government*. Cam-bridge, MA: Harvard University Press.

Renn, O., & Levine, D. (1991). Credibility and trust in risk communication. In R.E. Kasper-son and P.J.M. Stallen (Eds.), *Communicating Risks to the Public* (pp. 175–218). Dordrecht, The Netherlands: Kluwer Academic.

Slovic, P., Flynn, J., & Layman, M. (1991). Perceived risk, trust, and the politics of nuclear waste. *Science, 254*, 1603–1607.

6 Discussion and conclusion

The following presents a discussion of the research results that places the insights gained from the "Global Survey on Public Attitudes towards Climate Change" research project into the context of existing literature developed in the early chapters as well as with the underlying theories, hypothesis, and research questions of this study. The final part of this chapter addresses the meaning and implication of the knowledge gained for future communication efforts, emphasizes country-specific differences in the findings, and points to future research questions.

Public perceptions of climate change

The public is concerned about climate change, but considers it a low priority

The data show that on the global level, represented by the nine countries as a whole, the majority of the public expressed concern about climate change causing potential adverse impacts. In total, over 80 percent of the participants stated at least some concern regarding the possible impacts of climate change, and close to 32 percent indicated high levels of concern. This leads to the question, why are people concerned about climate change? Very strong levels of agreement were observed regarding the rationale that at some point, climate change will not be able to be reversed anymore. In total, more than three-quarters of the participants agreed or strongly agreed with that sentiment. In addition, almost two-thirds of the public seemed to be concerned due to the lack of political will to do something about it, and close to 50 percent are concerned because of the potential impacts of future generations (intergenerational equity).

However, the literature, especially for the United States, shows that climate change is considered a low priority in terms of political saliency. Studies show that the American public regards both the environment and climate change as relatively low national priorities (Bord et al., 1998; Dunlap & Scarce, 1991; Leiserowitz, 2010; Leiserowitz et al., 2010; Ockwell et al., 2009). The data collected in this study support the existing body of knowledge on the global scale and for different countries in regards to climate change as a low saliency issue. We found that on the global scale, climate change ranks last among societal issues that government needs to deal with. For the nine countries individually, climate change consistently ranks in the bottom third compared to higher priority issues such as increasing employment or improving education.

The low priority awarded to climate change on the global scale is reinforced by its perceived level of risk compared to other negative environmental impacts that can occur in society. Out of seven adverse environmental impacts such as worsening of urban air pollution, increasing frequency of droughts, or deterioration of the ozone layer, global climate change ranks fifth in terms of the public's perceived level of risk occurring over the next 20 years. Despite these comparatively low perceived risk levels, the public still wants the government to engage in efforts to reduce climate change and tends to be concerned about the lack of political will to do so.

Little difference between public risk perceptions of various adverse negative climate change impacts

Another aspect important for understanding the public's stand on climate change issue involves risk perceptions regarding climate change's negative impacts. On the global scale, the data do not show much variation in perceptions of different types of potential negative effects over the next 50 years. The survey participants were asked to rate the risk of climate change causing different harmful impacts. Respondents were given 13 consequences of climate change and were asked to evaluate their level of risk on a 5-point Likert scale. At least two-thirds of the surveyed populations believe that there are high or very high risks for climate change causing any of 13 harmful effects identified in the survey instrument. Nevertheless, responses in the "very high" risk category show that the public is most concerned about climate change causing "droughts and water shortage", "more frequent and serious floods", "forest fires", and "severe heat waves". On the other hand, only a small percentage of the public seems to believe that there is "no or little" risk of global climate change increasing the frequency or severity of environmental hazards in general.

Results indicate high levels of uncertainty

As pointed out in the literature (Renn & Levine, 1991; Kasperson et al., 1992; Nye Jr. et al., 1997), the public is more likely to support climate change policies if they trust the science behind it and the source of information. The data show that scientists are the most trusted source for climate change information, followed by television weather reports, family and friends, and environmental organizations. However, past studies in the Unites States also suggest that public climate change perceptions are influenced by various uncertainties that are critical in our understanding of climate change perceptions (Kempton, 1991; Seacrest et al., 2000; Leiserowitz, 2005; Lorenzoni et al., 2005; Smith, 2005;

Moser, 2006). This study identified several contradictions and hesitations among the surveyed population that indicate major uncertainties not only among the public in the United States, but also in other countries.

Although scientists are highly trusted and the most trusted source in all nine countries, the results also indicate that there is a large number of people who doubt the validity and sufficiency of the existing body of knowledge. The uncertainty is further emphasized by the high percentage of survey respondents who indicated that they are undecided on whether or not to trust any of the listed sources of climate change information in the survey instrument. Especially, in regards to teachers, 40 percent of all participants chose the answer category "undecided" followed by corporations (36 percent), family and friends (33.1 percent), and the mainstream news media (30.5 percent). In addition, the public perceives climate change as an issue that has to be solved to a large degree by the government and other institutions, but also ranks governmental organizations among the least trusted sources of climate change information. In terms of the public's level of support for adaptation and mitigation strategies, data show strong public support for efforts to reduce the causes and impacts of climate change. However, similar to the public's level of trust towards sources of climate change information, between a quarter and one-third of the people indicated that they are undecided to whether or not to support any climate change strategy. With climate change being still a controversial topic in the political arena and among some groups of the population, the people who are undecided today could make the difference in the future success of various climate change policies.

Public supports climate change policies in principle, but is less supportive of policies that directly affect them

Past studies show that the public largely supports policy action in general the national and international scale, but resists tax policies that directly affect it (Rosenstone et al., 1997; O'Connor et al., 1999; Moser, 2006; Leiserowitz et al., 2010). This study supports these findings on a global scale and for different countries. The results from the survey show ambiguity between the non-binding relatively strong support for adaptation and mitigation policies in general and the public's support for specific policies, willingness to pay more money for climate change abatement, and willingness to engage in behavioral changes to reduce climate change. The data show that less than a quarter of the total sample supports tax hikes as economic incentives to reduce the use of electricity or the use of automobiles. No more than one-third of the public supports tax hikes among any of the nine countries for mitigative policies.

Moreover, compared to past studies this research shows that the public not only refuses tax hikes but also hesitates to support any policies that require an initial investment on their part or changes to their daily routine or consumer practices. The analysis shows that the public strongly supports higher investments in public transit systems or transit-oriented developments, but more than one-third of the surveyed population also stated that they are not willing, or only slightly willing to change their travel behavior and use public transit systems more often. The hesitation is also supported by the fact that a relatively large number of survey participants stated that they are undecided in terms of their willingness to install solar panels or insulate their home, which would save money in the long run but require initial upfront capital investment.

Climate change is perceived as a general threat and not as a personal threat

The literature argues that for many people, especially in the developed world, climate change is still an issue removed in space and time, only effecting future generations in less developed countries (Bord et al., 1998; Lorenzoni et al., 2007). The recorded high levels of concern and the low ranking of climate change among other societal problems confirms and supports findings by other studies (Leiserowitz, 2005; Spence et al., 2012). The public does not attribute as much importance to climate change as to other, more tangible, visible, immediate, and urgent needs countries and governments face today. This may result from the long timeline perceived until impacts will be experienced.

Moreover, the data show that people recognize climate change as a high threat in general but not necessarily for themselves. Instead, high levels of threat are perceived predominantly for plants, animals, and people in other countries, which are rarely seen as personal threats. Thus, the percentage of people characterizing climate change as a high threat for people in their respective countries, to their family, and to themselves is significantly lower in all nine countries compared to the amount of people perceiving climate change as a high threat for people in other countries. When asked how long it will take until dangerous impacts of climate change will be experienced "somewhere on earth" or "in their region", the majority of the total study sample does believe that impacts are already occurring somewhere on earth, but not in their own region.

Perceptual factors and public support for climate change policies

The impact of socioeconomic variables

In order to identify potential factors that impact public's risk perception and policy support, one of the research objectives of this project was to see how socioeconomic characteristics impact the public's perceptions towards climate change. Therefore, regressions were used to test the following hypothesis: "The general attitude and public risk perceptions of climate change can be largely explained by socioeconomic variables". However, the regression analyses do not show any strong relationships between socioeconomic characteristics and climate change perceptions in any of the nine countries. Thus, the data collected do not confirm the hypothesis and show that characteristics such as age, gender, household income, or education are not strong predictors for someone's attitude or risk perceptions towards climate change. This is also seen in numerous studies in hazard research using social science methodology.

Perception factors impact public support for climate change policies and
willingness to commit to behavioral changes

Based on the third underlying hypothesis of this study, – the public's position towards climate change is the main reason for (1) low policy support,
(2) willingness to pay for climate change policies, and (3) willingness to change
their behavior as related to mitigation and adaptation – the impact of different
perceptual factors on public policy support were tested through regressions.
The first group of perceptual factors consisted of various risk perceptions of
climate change, and the second group included factors of trust, responsibility,
and knowledge related to climate change.

On the global scale, the data confirm a strong and significant relationship
between perception factors and the public's level of support for climate change
mitigation and adaptation policies. The regressions also show significant relationships between the perception factors and support for specific policies such
as energy efficiency strategies and urban planning strategies. In contrast, the
regressions do not show any strong links between climate change perceptions
and the public's willingness to pay more for climate change reduction among
any of the nine countries. Thus, even high perceived risk and threat levels, as
in the cases of Mexico, Brazil, or Germany, do not seem to impact the public's
level of disposition towards supporting climate change mitigation with more
private funds, but support for governmental policies in general. In contrast, in
terms of how well the perception factors can predict the public's willingness to
change their behavior towards a more sustainable lifestyle, the data established
strong relationships.

Perceived level of concern and personal responsibility have strongest
impact on policy support and willingness to change behavior

The literature does demonstrate that perceptions have significant impacts on
individual and group behavior and needs to be considered when developing
global climate change policies and strategies (Slovic, 2000). Furthermore, in
order to design, implement, and generate sufficient public support for policies
and planning interventions at the national and international level, it is necessary
to have a solid understanding of climate change perceptions (Read et al., 1994;
Bord et al., 1998; Moser, 2006; Moench, 2007).

Overall, the findings from regressions indicate that the predictor variables –
trust, responsibility, and knowledge – have stronger impact on the public's support for climate change policies compared to the different independent risk
perception variables – concern, attitude, and threat. However, to determine
which perceptual factors within the two groups has the strongest impact on
public behavior and policy support related to climate change, additional stepwise regressions were conducted. In stepwise regressions, the predictor variables
are entered into the model based on their statistical contribution in explaining
the variance in the dependent variable. Each time a predictor is added to the
equation, a removal test is made of the least useful predictor, thus identifying the

single independent variable that has the strongest relationship to the dependent variable. Among the different climate change risk perceptions captured by the survey instrument the stepwise regressions identified the "level of concern" over the impacts of climate change as the strongest of the independent or predictor variables for all countries. Varying by country, other perception factors with a strong impact on policy support are the perceived level of personal responsibility to reduce climate change and level of trust towards environmental organizations as a source of information.

Trust factors impact the general concern over climate change

As shown in Chapter 5, the public's level of concern over climate change impacts has the strongest effect on the level of support for mitigation and adaptation policies among all tested climate change risk perceptions. Typically, the more the public distrusts risk management and communicators information, the more concern they have about adverse impacts and potential threats for their own well-being (Slovic et al., 1991). However, much of this research was not conducted for hazards with such high levels of uncertainties as climate change. This led to the question of the role of trust factors impacting public concern in the context of climate change.

Regression results reveal that, on the global scale, approximately one-third of the variation in the public's level of concern can be explained by their level of trust towards climate change science and sources of information on climate change. Furthermore, the results show that the relationships between level of trust for the science and in different sources of information are not always positively correlated with level of concern. For example, the data indicate that the more people consider the climate change scientific data as trustworthy or trust environmental organizations, the more they are concerned about climate change. This is due to the fact that the scientific data and environmental organizations show major adverse or catastrophic impacts. In turn, the more the public trusts the information from corporations or governmental organizations, the less concerned they are about climate change. Although past studies often focused on technological risks, such as nuclear power (Bord & O'Connor, 1990; Slovic et al., 1991; Mushkatel & Pijawka, 1992), this study confirms similar relations for climate change trust and concern relationships.

Differences and similarities between the nine countries

Currently international studies on public risk perceptions are limited. Thus, very little is known about international perceptions on climate change threats and the perceptions that influence support for mitigation and adaptation policies (Leiserowitz, 2010; Schneider et al., 2010). Nevertheless, the existing literature does point out that perceptions are socially constructed and can vary by culture, human development, affluence, national experience with risks, and demographics (Slovic, 2010). Thus, we should expect widespread national differences.

Differences in climate change perceptions

Given that risk perceptions are culturally divided, a key hypothesis was that public perceptions of climate change in terms of threat and risk, saliency of the issue, trust in climate change information, and support for public mitigation strategies vary among countries. Approximately 90% of the populations of Mexico and Brazil seem to be concerned about climate change, perceive it as a high risk, and want their governments to take stronger action against the impacts and causes of climate change. In contrast, the Netherlands, United Kingdom, and the United States always place among the countries with the lowest amount of concern for climate change impacts, threats, and risks. In these three countries, only about 45% to 50% of the survey participant are concerned or highly concerned about the possible impacts of climate change. Furthermore, the populations of Mexico, Brazil, and Canada seem to be most willing to adjust their behavior to reduce climate change impacts. Still, the large number of people in all countries who indicated willingness to change behavior is very compelling.

In the United States and the Netherlands, over 24 percent of the people seem not at all willing to use public transit for most of their travel. This is significantly higher compared to the other seven surveyed countries. Japan, on the other hand, is last in terms of the willingness to purchase only energy saving appliances, install solar panels, or insulate their home or apartment. Together with Germany, Japan is also among the countries most skeptical regarding the sufficiency and trustworthiness of the scientific climate change data and the expert knowledge. Whereas, in the United States, over 10 percent, almost twice as many compared to any other of the nine countries, do not believe in the reality of climate change.

Despite these differences in climate change perceptions, the analysis also shows that some countries' perceptions are quite similar and the differences are only marginal. For example, Figure 6.1 shows a country's position with respect to the three principle perception factors – perceived personal level of threat, level of concern, and level of trust in the government's capability as risk manager in regards to global climate change. As illustrated by the scales of the X-, Y-, and Z-axis, the country differences among the mean scores of all three factors are small. Nevertheless, the differences are sufficient to cluster the countries into factor space. The data suggest that the United States, Germany, Japan, Spain, and the United Kingdom have more in common in terms of these critical risk perceptions than the differences. A second group consists of Mexico, Brazil, and Canada, which perceive the personal threat from climate change higher than any other countries. The Netherlands does not belong to any grouping. Although the Netherlands is very similar to the first group of countries in terms of personal concern and trust in government, the perceived level of personal threat from climate change is significantly lower compared to any other country. The low level of perceived personal threat among the Dutch public can be explained by the fact that the country is already strongly engaged

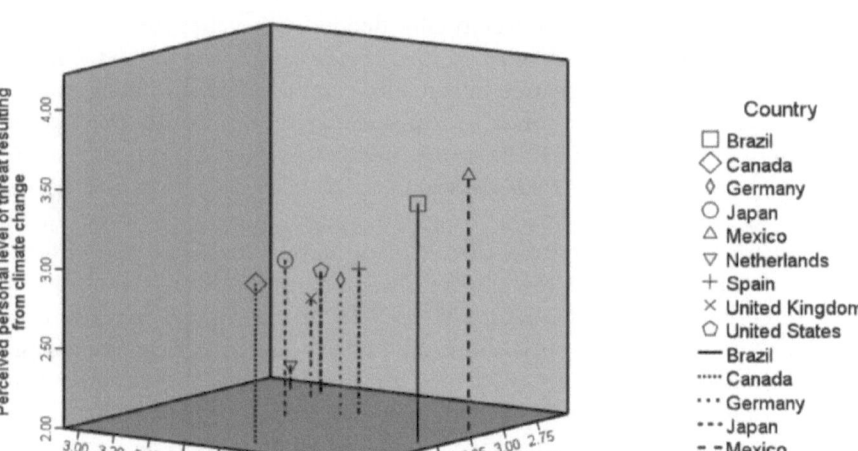

Figure 6.1 3D scatter plot of public climate change risk perceptions among the nine surveyed countries

Source: Author's illustration.

in adaptation measurements such as levies and flood gates due to its geographical circumstances. Most of the country is below sea level, and the public is used to living with the constant threat of floods and the resulting negative impacts.

The country groupings are different when based on the public's perceived level of personal responsibility, knowledge, and overall trust towards climate change information sources. This can be viewed as the "trust factor". As illustrated in Figure 6.2, the countries can be divided into four groups. Brazil and Mexico have the highest mean scores for all three factors and thus can be clustered. The second group consists of Spain, United States, Japan, and Canada, which are very close together in terms of perceived level of knowledge as well as personal responsibility to reduce climate change and only vary slightly regarding the general level of trust towards different sources of climate change information. The United Kingdom and Germany can be grouped together based on their sense of personal responsibility and general climate change knowledge. In addition, the difference in the mean scores for overall trust in climate change information is only 0.11 between the two countries. Again, the Netherlands is the outlier with a low perceived personal responsibility to reduce climate change compared to the eight other countries.

With respect to the level of support for adaptation and mitigation policies and the willingness to commit to behavioral changes, Figure 6.3 shows the positioning of the nine countries. Again, Mexico and Brazil are the most supportive of climate change policies in general as well as the most willing to

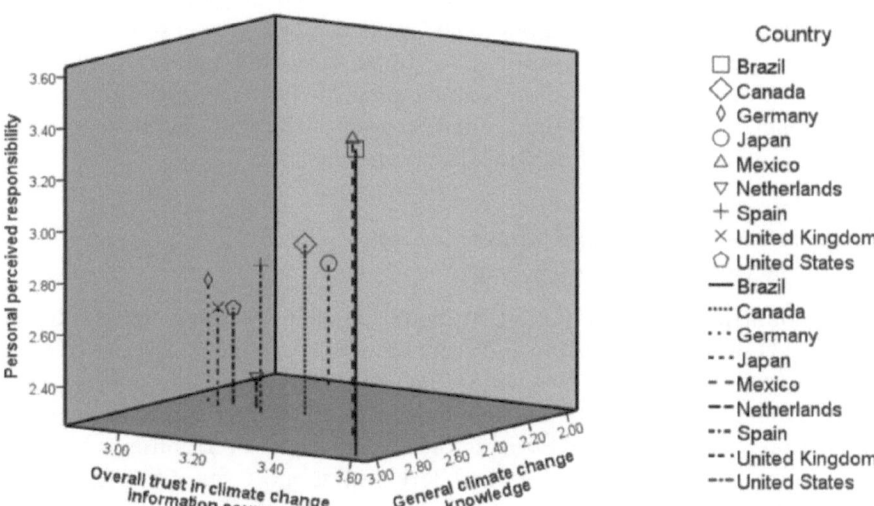

Figure 6.2 3D scatter plot of public perceptions of factors of trust, knowledge, and responsibility

Source: Author's illustration.

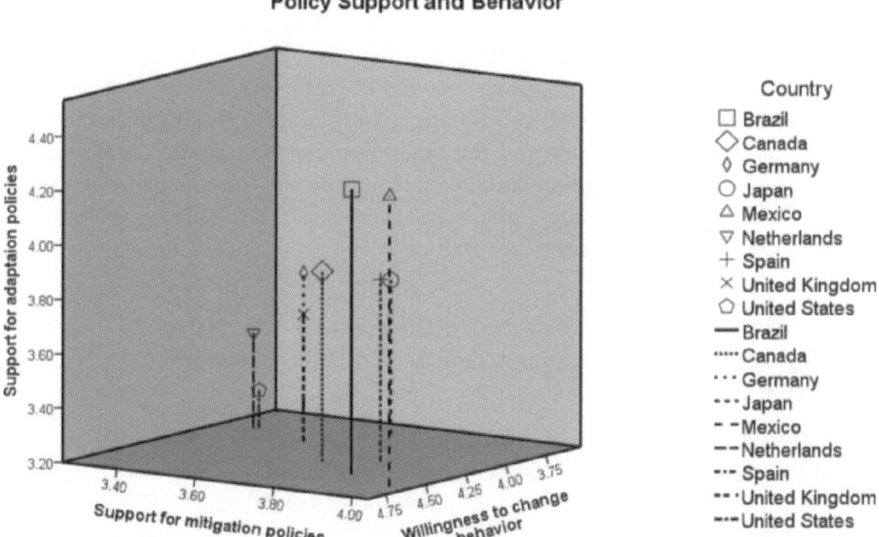

Figure 6.3 3D scatter plot of public climate change policy support and willingness to change behavior

Source: Author's illustration.

change behavior to reduce the causes of climate change. Data from Germany, Canada, Spain, and Japan show strong similarities regarding support for mitigation policies as well as willingness to change behavior and only show slight differences in support for mitigation. The United Kingdom and the Netherlands are very close in terms of behavior and among the least supportive for adaptation policies. The outlier is the United States for which the public is the least supportive of both mitigation and adaptation policies.

Differences in the relationships between perception factors and climate change policy support

Most countries in this study show strong relationships between climate change risk perceptions and mitigation and adaptation policy support. In terms of more specific policies grouped by themes, such as energy efficiency policies, economic incentives, and planning and adaptation strategies, the country-specific regression results were mixed. That is, the strengths of the relationships between the different perception factors and levels of support for different climate change strategies are not identical among the countries. The data suggest that the relationships are the strongest among the participants from the United States, followed by the United Kingdom, Spain, Netherlands, Germany, Canada, and Japan. In the cases of Brazil and Mexico, neither attitude, levels of concern nor factors of trust, responsibility, and knowledge seem to explain any variability in the public's support for or opposition to mitigation and adaptation policies. This is not necessarily surprising, considering that, compared to the other highly industrialized countries in our survey, the perceived levels of risk or concerns regarding climate change are higher, and more people believe that impacts are already occurring in their country.

In terms of the impact of knowledge, as well as factors of trust and responsibility on the public's support for climate change policies, the regressions show significant relationships among all the nine countries. The results show a strong relationship between the independent variables and the public's general support for mitigation and adaptation policies in seven out of the nine countries, with Mexico and Brazil the exceptions again.

Contributions to the underlying theories of the "Global Survey on Public Attitudes towards Climate Change" research project

As discussed in Chapter 2, the research was based on the psychometric paradigm (Fischhoff et al., 1978; Slovic et al., 1984). Since past perceptual research focused primarily on technological risks and natural hazards (Burton et al., 1978; Kates, 1982; Short, 1984; Slovic, 2000), which are quite different from climate change, we do not know very much about the relationship between the different perception factors in the context of climate change. By testing the role of heuristics, trust, values (or worldviews), and social amplification related to climate change risk perceptions among nine identified countries this research contributes to the fundamental theories of "bounded rationality" (Simon, 1956, 1959), and "cultural cognition" of natural phenomenon (Douglas & Wildavsky, 1982) from a global climate change perceptive.

In the context of climate change, the data indicate that the public applies invalid heuristics resulting in risk assessments and levels of policy support contrary to the scientific findings and recommendations provided by the scientific community. The study shows that the public underrates the personal risks of climate change, compared to scientists who point out that, even in industrialized countries like the United States, climate change is already occurring and posing dangerous impacts (Pittock, 2009). As discussed earlier, the majority of the public believes that climate change is an issue removed in time and space that will primarily impact future generations in other countries. This shows that the public does not have sufficient information or the cognitive skills necessary to make well-informed decisions that would confirm the conclusions by the scientific community. Therefore, this study confirms the concept of heuristics and the argument that if heurists are invalid for the risk faced they can lead to or reinforce existing misconceptions (Kahneman et al., 1982; Makofske & Edelstein, 1988). One important reason for this gap between lay models and expert assessment is the public's skepticism and uncertainty of the existing body of climate change knowledge and the low levels of trust towards key risk communicators such as the media and governmental organizations. Furthermore, the low raking of climate change compared to other societal issues and other environmental impacts further dampens the public's missing sense of urgency to take action against climate change. As a result, effective policies to reduce greenhouse gas emissions such as higher taxes related to energy use and behavioral changes to travel behavior are the least supported strategies by the public.

Another theory investigated is that lay people dealing with uncertainties tend to over- or underestimate the risks and threats (Slovic, 2000). In the case of nuclear power risk perception, research shows that the public tends to overrate the risks of radiation exposure leading to large social amplification effects and behavior. This study advances the theory of social amplification by examining whether or not current forms of climate change communication amplify or attenuate the public's risk perception and by discussing the potential of social amplification for climate change compared to other environmental hazards.

The low political saliency and low ranking of climate change compared to other environmental hazards suggests that communication efforts have not amplified the perceptions of risk and their manageability. Instead, the data and literature suggest current communication of climate change, especially by the mass media, has attenuated the public's perceived urgency of dealing with the issue of climate change. According to the data, the majority of the publics do not believe that the media's attention on the effects of climate change is exaggerated which otherwise could lead to an increased risk perception, nor is the media considered a source for climate change information that can be strongly trusted. All current major communicators such as governmental institutions or scientist are either not trusted by the public or not capable to convey the risks and impacts of climate change in a convincing and easy to understand manner (Sheppard, 2012).

Thus, this research suggests that current forms of climate change communication are not very likely to amplify the public's risk perception. However, this might change when impacts of climate change become more obvious to the public, levels of trust towards communicators increase, and reporting links highly publicized negative events, such as droughts, hurricanes, or floods more often to global climate change. Existing research also shows a negative relationship between risk or threat perception and trust factors, suggesting that as perceived risk increases as the level of institutional trust decreases (Slovic et al., 1991). Being important, this led to the question if such relationship between trust factors and public risk perceptions also exist in the context of climate change. Therefore, the theory tested in the context of climate change is that distrust of the government is strongly linked to the level of risk perceived.

In the context of climate change, the data only show a moderate relationship at best between the public's climate change threat and risk perceptions and their level of trust towards the government. Nevertheless, the study confirms the relationship between trust in government and level of risk perceived in the context of climate change. Moreover, the analysis shows that the stronger the public believes in climate change risk and threat the more they trust the government's capability as climate change risk manager. This marks a significant difference compared to the negative relationships between risk or threat perceptions and trust in government for other risks, such as nuclear power. Therefore, the results show that the uniqueness of the climate change issue with its high uncertainties, mostly invisible causes and slow developing impacts also changes the typical type of relationship between risk and treat perceptions and institutional trust.

Cultural theorists argue that our worldviews and our values play an important role in public risk perception and behavior (Doulas, 1966, 1970; Douglas & Wildavsky, 1982; Douglas et al., 1998). This study provided empirical evidence to support the argument that we disagree about climate change because we have different belief systems mediated through culture (Hulme, 2009; Kahan et al., 2011) and thus contributes to the theory of cultural cognition. The previously discussed perceptual differences among the nine surveyed countries suggest cultural differences among the survey participants, which in turn impact the perceptual factors and behavior. Furthermore, the data show characteristics of intergeneration equity among a large number of participants. Moreover, variables capturing the concern for family members, as well as perceived personal responsibility for reducing climate change, are confirmed by stepwise regressions to have a significant impact on policy support. Thus, the collected data further confirm to the theory of cultural cognition by showing that cultural background and personal values play a role in public risk perception and behavior in the context of climate change.

Implications for climate change communication programs

Communication efforts can foster a personal connection to climate change, raise the level of concern, and thus increase the level of support for mitigation and adaptation policies as well as the willingness among the public to engage in a more sustainable behavior. This study has identified several aspects that need to be considered in future communication programs. Uncertainty is seen throughout the survey questions, suggesting the critical importance of risk communication programs. A large number of people are uncertain about the danger climate change poses today or for future generations, do not know which source of information to trust, resulting in indecision to whether or not support mitigation or adaptation strategies.

However, the high level of uncertainty among the public also presents an opportunity to increase policy support and foster behavioral changes in the future through well-designed communication programs. With climate change being still a controversial topic in the political arena and among some groups of the population, the people who are undecided today could very well make the difference in the future success of various climate change policies. The comparatively high levels of public uncertainty and indecisiveness show that public behavior and perceptions can still be influenced by objective climate change coverage if they establish personal connections to climate change impacts, which is likely to increase the level of concern and support for mitigation and adaptation policies.

The study also shows that, perceptual factors of trust and responsibility have a great impact on public behavior and policy support and thus need to be acknowledged in communication efforts. The level of success of risk communication is significantly influenced by the public's trust in the communicator and in the ability of individuals, industries, or institutions responsible for risk management. Trust in organizations whose risk management policies impact communities and the environment is vital in order to reduce complexity and generate social cooperation (Cvetkovich & Loefstedt, 1999). On the one hand, perceptions of trust have a strong impact on public concern which in turn influences public support for climate change mitigation and adaptation. On the other hand, the results also show that trust also directly influence public behavior and policy support. With respect to the public's level of trust towards different sources of information, scientists seem to be most trusted among potential communicators in all nine countries, followed by family and friends. Analyses also show that the level of trust specifically in environmental organizations and perceived trustworthiness of climate change science have the strongest impacts on perceived levels of concern.

However, data also show significant barriers to successful communication efforts. The most significant barrier is probably the fact that while the public views the government as the party most responsible to reduce climate change, they are simultaneously highly distrusting of it. Communication efforts need to acknowledge this contradiction to build up trust and motivate the public to be more engaged in reducing climate change. This might be accomplished by emphasizing the multiple benefits of many policies aimed at more than just reducing climate change.

Another obstacle is trust in the data. Over one-third of the participants doubt that the scientific community actually has enough data to fully understand the complexity of the issue. This may explain the widespread skepticism regarding the trustworthiness of climate change findings as well. Levels of skepticism towards the trustworthiness and sufficiency of scientific findings as well as the reality of climate change vary by country. Thus, communicators need to be aware of their audience in order to decide how educational their program needs to be. Climate change risk perceptions, levels of policy support, and the interrelationships between the two need to be constantly reevaluated in order to improve communication programs and decrease the gap between scientific community recommendations and public and policy makers' actions.

Future research

As evident from the analytical framework and the feedback loop in particular, global climate change risk perceptions, levels of policy support, and their interrelationships need to be constantly reevaluated in order to improve communication programs and to decrease the gap between the recommendations provided by the scientific community and the actual actions by the public and policy makers. Therefore, this study should function as a benchmark for various follow-up studies, adding more countries to the database as well as enabling longitudinal research for the countries addressed in this study. In addition, research with larger sample sizes per country and more survey questions is needed to further improve the understanding of the perceptual differences between countries and what variables can explain them. To improve climate change communication programs, future research should also incorporate interviews of public officials directly involved with past or ongoing climate change communication efforts.

Bibliography

Bord, R.J., Fisher, A., & O'Connor, R.E. (1998). Public perceptions of global warming: United States and international perspectives. *Climate Research, 11,* 75–84.

Bord, R.J., & O'Connor, R.E. (1990). Risk communication, knowledge, and attitudes: Explaining reactions to a technology perceived as risky. *Risk Analysis, 10,* 499–506.

Burton, I., Kates, R.W., & White, G.F. (1978). *The Environment as Hazard.* New York, NY: Oxford University Press.

Cvetkovich, G., & Loefstedt, R. (Eds.). (1999). *Social Trust and the Management of Risk.* London, UK: Earthscan.

Douglas, M. (1966). *Purity and Danger: An Analysis of Concepts of Pollution and Taboo.* London, UK: Taylor

Douglas, M. (1970). *Natural Symbols: Explorations in Cosmology.* London, UK: Barrie and Rockliff.

Douglas, M., Gasper, D., Ney, S., & Thompson, M. (1998). Human needs and wants. In S. Rayner & E.L. Malone (Eds.), *Human Choice and Climate Change* (pp. 195–265). Columbus, OH: Battelle Press.

Douglas, M., & Wildavsky, A. (1982). *Risk and Culture: An Essay on the Selection of Technological and Environmental Dangers*. Berkeley: University of California Press.

Dunlap, R.E., & Scarce, R. (1991). The polls – poll trends: Environment problems and protection. *Public Opinion Quarterly, 55*, 651–672.

Fischhoff, B., Slovic, P., Lichtenstein, S., Read, S., & Combs, B. (1978). How safe is safe enough? A psychometric study of attitudes towards technological risks and benefits. *Policy Sciences, 9*, 127–152.

Hulme, M. (2009). *Why We Disagree about Climate Change: Understanding Controversy Inaction and Opportunity*. New York, NY: Cambridge University Press.

Kahan, D.M., Peters, E., Braman, D., Slovic, P., Wittlin, M., Ouellette, L.L., & Mandel, G. (2011). The Tragedy of the Risk-Perception Commons: Culture Conflict, Rationality Conflict, and Climate Change Cultural Cognition Project. *Cultural Cognition Project, Working Paper No. 89*.

Kahneman, D., Slovic, P., & Tversky, A. (Eds.). (1982). *Judgment under Uncertainty: Heuristics and Biases*. New York, NY: Cambridge University Press.

Kasperson, R., Golding, D., & Tuler, S. (1992). Social distrust as a factor in sitting hazardous facilities and communicating risks. *Journal of Social Issues, 48*(4), 161–187.

Kates, R.W. (1982). *Risk Assessment of Environmental Hazards*. Chichester, UK: Wiley.

Kempton, W. (1991). Lay perspectives on global climate change. *Global Environmental Change, 1*(3), 183–208.

Leiserowitz, A.A. (2005). American risk perceptions: Is climate change dangerous? *Risk Analysis, 25*(6), 1433–1442.

Leiserowitz, A. (2010). Risk perception and behavior. In S.H. Schneider, A. Rosencranz, M.D. Mastrandrea, & K. Kuntz-Duriseti (Eds.), *Climate Change Science and Policy* (pp. 175–184). Washington, DC: Island Press.

Leiserowitz, A., Maibach, E., & Roser-Renouf, C. (2010). *Climate Change in the American Mind: Americans' Global Warming Beliefs and Attitudes in January 2010*. New Haven, CT: Yale University and Mason University.

Lorenzoni, I., Pidgeon, N., & O'Connor, R. (2005). Dangerous climate change: The role for risk research. *Risk Analysis, 25*(6), 1387–1398.

Lorenzoni, I., Nicholson-Cole, S., & Whitmarsh, L. (2007). Barriers perceived to engaging with climate change among the UK public and their policy implications. *Global Environmental Change, 17*(3–4), 445–459.

Makofske, J.W., & Edelstein, M.R. (1988). *Radon and the Environment*. Park Ridge, IL: Elsevier Science.

Moench, M. (2007). Adapting to climate change and the risks associated with other natural hazards: Methods for moving from concepts to action. In M. Moench & A. Dixit (Eds.), *Working with the Winds of Change*. Kathmandu, Nepal: ISET-Nepal.

Moser, S.C. (2006). Talk of the city: Engaging urbanities on climate change. *Environmental Research Letters, 1*, 1–10.

Mushkatel, A.H., & Pijawka, K.D. (1992). *Institutional Trust, Information and Risk Perceptions: Report of Findings of the Las Vegas Metropolitan Area Survey*. Carson City: Nevada Nuclear Waste Project Office.

Nye Jr., J.S., Zelikow, P.D., & King, D.C. (1997). *Why People Don't Trust Government*. Cambridge, MA: Harvard University Press.

Ockwell, D., Whitmarsh, L., & O'Neill, S. (2009). Reorienting climate change communication for effective mitigation: Forcing people to be green or fostering grass-roots engagement? *Science Communication, 30*(3), 305–327.

O'Connor, R., Bard, R., & Fisher, A. (1999). Risk perceptions, general environmental beliefs, and willingness to address climate change. *Risk Analysis, 19*(3), 461–471.

Pittock, B.A. 2009. *Climate Change: The Science, Impacts and Solutions*. London, UK: Earthscan.

Read, D., Bostrom, A., Granger Morgan, M., Fischhoff, B., & Smuts, T. (1994). What do people know about global climate change? 2. Survey Studies of educated laypeople. *Risk Analysis, 14*(6), 971–982.

Renn, O., & Levine, D. (1991). Credibility and trust in risk communication. In R.E. Kasperson and P.J.M. Stallen (Eds.), *Communicating Risks to the Public* (pp. 175–218). Dordrecht, The Netherlands: Kluwer Academic.

Rosenstone, S., Kinde, D., & Miller, W. (1997). *American National Election Study*. Ann Arbor, MI: Inter-university Consortium for Political and Social Research.

Schneider, S.H., Rosencranz, A., Mastrandrea, M.D., & Kuntz-Duriseti, K. (Eds.). (2010). *Climate Change Science and Policy*. Washington, DC: Island Press.

Seacrest, S., Kuzelka, R., & Rick, L. (2000). Global Climate Change and Public Perception: The Challenge of Translation. *Journal of the American Water Resources Association, 36*(2), 253–263.

Sheppard, R.J. (2012). *Visualizing Climate Change: A Guide to Visual Communication of Climate Change and Developing Local Solutions*. New York, NY: Routledge.

Short, J.F. (1984). The social fabric at risk: Toward the social transformation of risk analysis. *American Sociological Review, 49*, 711–725.

Simon, H.A. (1956). Rational choice and the structure of the environment. *Psychological Review, 63*, 129–138.

Simon, H.A. (1959). Theories of decision making in economics and behavioral science. *American Economic Review, 49*, 253–283.

Slovic, P. (2000). *The Perception of Risk*. London, UK: Earthscan.

Slovic. P. (2010). *The Feeling of Risk*. London, UK: Earthscan.

Slovic, P., Fischhoff, B., & Lichtenstein, S. (1984). Behavioral decision theory perspectives on risk and safety. *Acta Psychologica, 56*, 183–203.

Slovic, P., Flynn, J., & Layman, M. (1991). Perceived risk, trust, and the politics of nuclear waste. *Science, 254*, 1603–1607.

Smith, J. (2005). Dangerous news: Media decision making about climate change risk. *Risk Analysis, 25*(6), 1471–1482.

Spence, A., Pootinage, W., & Pidgeon, N. 2012. The psychological distance of climate change. *Risk Analyses, 32*(6), 957–971.

Index

Note: Page numbers with *f* indicate figures; those with *t* indicate tables.